The Meson

Edited by Paul F. Kisak

Contents

Chapter 1

Meson

In particle physics, **mesons** (/ˈmiːzɒnz/ or /ˈmɛzɒnz/) are hadronic subatomic particles composed of one quark and one antiquark, bound together by the strong interaction. Because mesons are composed of sub-particles, they have a physical size, with a diameter of roughly one fermi, which is about $2/3$ the size of a proton or neutron. All mesons are unstable, with the longest-lived lasting for only a few hundredths of a microsecond. Charged mesons decay (sometimes through intermediate particles) to form electrons and neutrinos. Uncharged mesons may decay to photons.

Mesons are not produced by radioactive decay, but appear in nature only as short-lived products of very high-energy interactions in matter, between particles made of quarks. In cosmic ray interactions, for example, such particles are ordinary protons and neutrons. Mesons are also frequently produced artificially in high-energy particle accelerators that collide protons, anti-protons, or other particles.

In nature, the importance of lighter mesons is that they are the associated quantum-field particles that transmit the nuclear force, in the same way that photons are the particles that transmit the electromagnetic force. The higher energy (more massive) mesons were created momentarily in the Big Bang, but are not thought to play a role in nature today. However, such particles are regularly created in experiments, in order to understand the nature of the heavier types of quark that compose the heavier mesons.

Mesons are part of the hadron particle family, defined simply as particles composed of two quarks. The other members of the hadron family are the baryons: subatomic particles composed of three quarks rather than two. Some experiments show evidence of exotic mesons, which don't have the conventional valence quark content of one quark and one antiquark.

Because quarks have a spin of $1/2$, the difference in quark-number between mesons and baryons results in conventional two-quark mesons being bosons, whereas baryons are fermions.

Each type of meson has a corresponding antiparticle (antimeson) in which quarks are replaced by their corresponding antiquarks and vice versa. For example, a positive pion (π+) is made of one up quark and one down antiquark; and its corresponding antiparticle, the negative pion (π−), is made of one up antiquark and one down quark.

Because mesons are composed of quarks, they participate in both the weak and strong interactions. Mesons with net electric charge also participate in the electromagnetic interaction. They are classified according to their quark content, total angular momentum, parity and various other properties, such as C-parity and G-parity. Although no meson is stable, those of lower mass are nonetheless more stable than the most massive mesons, and are easier to observe and study in particle accelerators or in cosmic ray experiments. They are also typically less massive than baryons, meaning that they are more easily produced in experiments, and thus exhibit certain higher energy phenomena more readily than baryons composed of the same quarks would. For example, the charm quark was first seen in the J/Psi meson (J/ψ) in 1974,[1][2] and the bottom quark in the upsilon meson (ϒ) in 1977.[3]

1.1 History

From theoretical considerations, in 1934 Hideki Yukawa[4][5] predicted the existence and the approximate mass of the "meson" as the carrier of the nuclear force that holds atomic nuclei together. If there were no nuclear force, all nuclei with two or more protons would fly apart because of the electromagnetic repulsion. Yukawa called his carrier particle the meson, from μέσος *mesos*, the Greek word for "intermediate," because its predicted mass was between that of the electron and that of the proton, which has about 1,836 times the mass of the electron. Yukawa had originally named his particle the "mesotron", but he was corrected by the physicist Werner Heisenberg (whose father was a professor of Greek at the University of Munich). Heisenberg pointed out that there is no "tr" in the Greek word "mesos".[6]

The first candidate for Yukawa's meson, now known in modern terminology as the muon, was discovered in 1936 by Carl David Anderson and others in the decay products of cosmic ray interactions. The mu meson had about the right mass to be Yukawa's carrier of the strong nuclear force, but over the course of the next decade, it became evident that it was not the right particle. It was eventually found that the "mu meson" did not participate in the strong nuclear interaction at all, but rather behaved like a heavy version of the electron, and was eventually classed as a lepton like the electron, rather than a meson. Physicists in making this choice decided that properties other than particle mass should control their classification.

There were years of delays in the subatomic particle research during World War II in 1939–45, with most physicists working in applied projects for wartime necessities. When the war ended in August 1945, many physicists gradually returned to peacetime research. The first true meson to be discovered was what would later be called the "pi meson" (or pion). This discovery was made in 1947, by Cecil Powell, César Lattes, and Giuseppe Occhialini, who were investigating cosmic ray products at the University of Bristol in England, based on photographic films placed in the Andes mountains. Some mesons in these films had about the same mass as the already-known meson, yet seemed to decay into it, leading physicist Robert Marshak to hypothesize in 1947 that it was actually a new and different meson. Over the next few years, more experiments showed that the pion was indeed involved in strong interactions. The pion (as a virtual particle) is the primary force carrier for the nuclear force in atomic nuclei. Other mesons, such as the rho mesons are involved in mediating this force as well, but to lesser extents. Following the discovery of the pion, Yukawa was awarded the 1949 Nobel Prize in Physics for his predictions.

The word *meson* has at times been used to mean *any* force carrier, such as the "Z^0 meson", which is involved in mediating the weak interaction.[7] However, this spurious usage has fallen out of favor. Mesons are now defined as particles composed of pairs of quarks and antiquarks.

1.2 Overview

1.2.1 Spin, orbital angular momentum, and total angular momentum

Main articles: Spin (physics), angular momentum operator, Total angular momentum and Quantum numbers

Spin (quantum number S) is a vector quantity that represents the "intrinsic" angular momentum of a particle. It comes in increments of $\frac{1}{2}$ ħ. The ħ is often dropped because it is the "fundamental" unit of spin, and it is implied that "spin 1" means "spin 1 ħ". (In some systems of natural units, ħ is chosen to be 1, and therefore does not appear in equations).

Quarks are fermions—specifically in this case, particles having spin $\frac{1}{2}$ ($S = \frac{1}{2}$). Because spin projections vary in increments of 1 (that is 1 ħ), a single quark has a spin vector of length $\frac{1}{2}$, and has two spin projections ($S_z = +\frac{1}{2}$ and $S_z = -\frac{1}{2}$). Two quarks can have their spins aligned, in which case the two spin vectors add to make a vector of length $S = 1$ and three spin projections ($S_z = +1$, $S_z = 0$, and $S_z = -1$), called the spin-1 triplet. If two quarks have unaligned spins, the spin vectors add up to make a vector of length S = 0 and only one spin projection ($S_z = 0$), called the spin-0 singlet. Because mesons are made of one quark and one antiquark, they can be found in triplet and singlet spin states.

There is another quantity of quantized angular momentum, called the orbital angular momentum (quantum number L), that comes in increments of 1 ħ, which represent the angular momentum due to quarks orbiting around each other. The total angular momentum (quantum number J) of a particle is therefore the combination of intrinsic angular momentum (spin) and orbital angular momentum. It can take any value from $J = |L - S|$ to $J = |L + S|$, in increments of 1.

Particle physicists are most interested in mesons with no orbital angular momentum ($L = 0$), therefore the two groups of mesons most studied are the $S = 1$; $L = 0$ and $S = 0$; $L = 0$, which corresponds to $J = 1$ and $J = 0$, although they are not the only ones. It is also possible to obtain $J = 1$ particles from $S = 0$ and $L = 1$. How to distinguish between the $S = 1$, $L = 0$ and $S = 0$, $L = 1$ mesons is an active area of research in meson spectroscopy.

1.2.2 Parity

Main article: Parity (physics)

If the universe were reflected in a mirror, most of the laws of physics would be identical—things would behave the same way regardless of what we call "left" and what we call "right". This concept of mirror reflection is called parity (P). Gravity, the electromagnetic force, and the strong interaction all behave in the same way regardless of whether or not the universe is reflected in a mirror, and thus are said to conserve parity (P-symmetry). However, the weak interaction does distinguish "left" from "right", a phenomenon called parity violation (P-violation).

Based on this, one might think that, if the wavefunction for each particle (more precisely, the quantum field for each particle type) were simultaneously mirror-reversed, then the new set of wavefunctions would perfectly satisfy the laws of physics (apart from the weak interaction). It turns out that this is not quite true: In order for the equations to be satisfied, the wavefunctions of certain types of particles have to be multiplied by −1, in addition to being mirror-reversed. Such particle types are said to have *negative* or *odd* parity ($P = -1$, or alternatively $P = -$), whereas the other particles are said to have *positive* or *even* parity ($P = +1$, or alternatively $P = +$).

For mesons, the parity is related to the orbital angular momentum by the relation:[8]

$$P = (-1)^{L+1}$$

where the L is a result of the parity of the corresponding spherical harmonic of the wavefunction. The '+1' in the exponent comes from the fact that, according to the Dirac equation, a quark and an antiquark have opposite intrinsic parities. Therefore, the intrinsic parity of a meson is the product of the intrinsic parities of the quark (+1) and antiquark (−1). As these are different, their product is −1, and so it contributes a +1 in the exponent.

As a consequence, mesons with no orbital angular momentum ($L = 0$) all have odd parity ($P = -1$).

1.2.3 C-parity

Main article: C-parity

C-parity is only defined for mesons that are their own antiparticle (i.e. neutral mesons). It represents whether or not the wavefunction of the meson remains the same under the interchange of their quark with their antiquark.[9] If

$$|q\bar{q}\rangle = |\bar{q}q\rangle$$

then, the meson is "C even" (C = +1). On the other hand, if

$$|q\bar{q}\rangle = -|\bar{q}q\rangle$$

then the meson is "C odd" (C = −1).

C-parity rarely is studied on its own, but more commonly in combination with P-parity into CP-parity. CP-parity was thought to be conserved, but was later found to be violated in weak interactions.[10][11][12]

1.2.4   G-parity

Main article: G-parity

G parity is a generalization of the C-parity. Instead of simply comparing the wavefunction after exchanging quarks and antiquarks, it compares the wavefunction after exchanging the meson for the corresponding antimeson, regardless of quark content.[13] In the case of neutral meson, G-parity is equivalent to C-parity because neutral mesons are their own antiparticles.

If

$$|q_1\bar{q}_2\rangle = |\bar{q}_1 q_2\rangle$$

then, the meson is "G even" (G = +1). On the other hand, if

$$|q_1\bar{q}_2\rangle = -|\bar{q}_1 q_2\rangle$$

then the meson is "G odd" (G = −1).

1.2.5   Isospin and charge

Main article: Isospin

The concept of isospin was first proposed by Werner Heisenberg in 1932 to explain the similarities between protons and neutrons under the strong interaction.[14] Although they had different electric charges, their masses were so similar that physicists believed that they were actually the same particle. The different electric charges were explained as being the result of some unknown excitation similar to spin. This unknown excitation was later dubbed *isospin* by Eugene Wigner in 1937.[15] When the first mesons were discovered, they too were seen through the eyes of isospin and so the three pions were believed to be the same particle, but in different isospin states.

This belief lasted until Murray Gell-Mann proposed the quark model in 1964 (containing originally only the u, d, and s quarks).[16] The success of the isospin model is now understood to be the result of the similar masses of the u and d quarks. Because the u and d quarks have similar masses, particles made of the same number of them also have similar masses. The exact specific u and d quark composition determines the charge, because u quarks carry charge $+2/3$ whereas d quarks carry charge $-1/3$. For example the three pions all have different charges ($\pi+$ (ud), $\pi 0$ (a quantum superposition of uu and dd states), $\pi-$ (du)), but have similar masses (~140 MeV/c^2) as they are each made of a same number of total of up and down quarks and antiquarks. Under the isospin model, they were considered to be a single particle in different charged states.

The mathematics of isospin was modeled after that of spin. Isospin projections varied in increments of 1 just like those of spin, and to each projection was associated a "charged state". Because the "pion particle" had three "charged states", it was said to be of isospin $I = 1$. Its "charged states" $\pi+$, $\pi 0$, and $\pi-$, corresponded to the isospin projections $I_3 = +1$, $I_3 = 0$, and $I_3 = -1$ respectively. Another example is the "rho particle", also with three charged states. Its "charged states" $\rho+$, $\rho 0$, and $\rho-$, corresponded to the isospin projections $I_3 = +1$, $I_3 = 0$, and $I_3 = -1$ respectively. It was later noted that the isospin projections were related to the up and down quark content of particles by the relation

$$I_3 = \frac{1}{2}[(n_u - n_{\bar{u}}) - (n_d - n_{\bar{d}})],$$

where the n's are the number of up and down quarks and antiquarks.

In the "isospin picture", the three pions and three rhos were thought to be the different states of two particles. However, in the quark model, the rhos are excited states of pions. Isospin, although conveying an inaccurate picture of things, is still used to classify hadrons, leading to unnatural and often confusing nomenclature. Because mesons are hadrons, the isospin classification is also used, with $I_3 = +1/2$ for up quarks and down antiquarks, and $I_3 = -1/2$ for up antiquarks and down quarks.

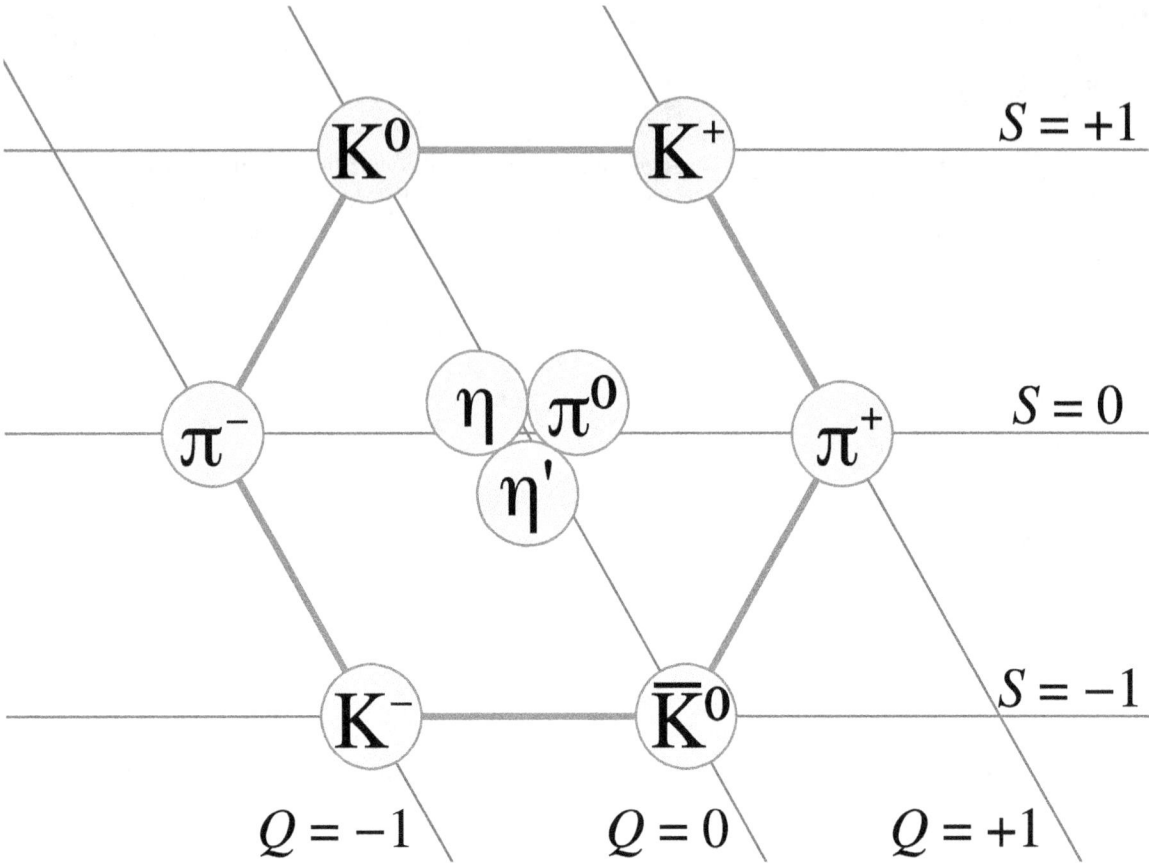

Combinations of one u, d or s quarks and one u, d, or s antiquark in $J^P = 0^-$ configuration form a nonet.

1.2.6 Flavour quantum numbers

Main article: Flavour (particle physics) § Flavour quantum numbers

The strangeness quantum number S (not to be confused with spin) was noticed to go up and down along with particle mass. The higher the mass, the lower the strangeness (the more s quarks). Particles could be described with isospin projections (related to charge) and strangeness (mass) (see the uds nonet figures). As other quarks were discovered, new quantum numbers were made to have similar description of udc and udb nonets. Because only the u and d mass are similar, this description of particle mass and charge in terms of isospin and flavour quantum numbers only works well for the nonets made of one u, one d and one other quark and breaks down for the other nonets (for example ucb nonet). If the quarks all had the same mass, their behaviour would be called *symmetric*, because they would all behave in exactly the same way with respect to the strong interaction. However, as quarks do not have the same mass, they do not interact in the same way (exactly like an electron placed in an electric field will accelerate more than a proton placed in the same field because of its lighter mass), and the symmetry is said to be broken.

It was noted that charge (Q) was related to the isospin projection (I_3), the baryon number (B) and flavour quantum numbers (S, C, B', T) by the Gell-Mann–Nishijima formula:[17]

$$Q = I_3 + \frac{1}{2}(B + S + C + B' + T),$$

where $S, C, B',$ and T represent the strangeness, charm, bottomness and topness flavour quantum numbers respectively. They are related to the number of strange, charm, bottom, and top quarks and antiquark according to the relations:

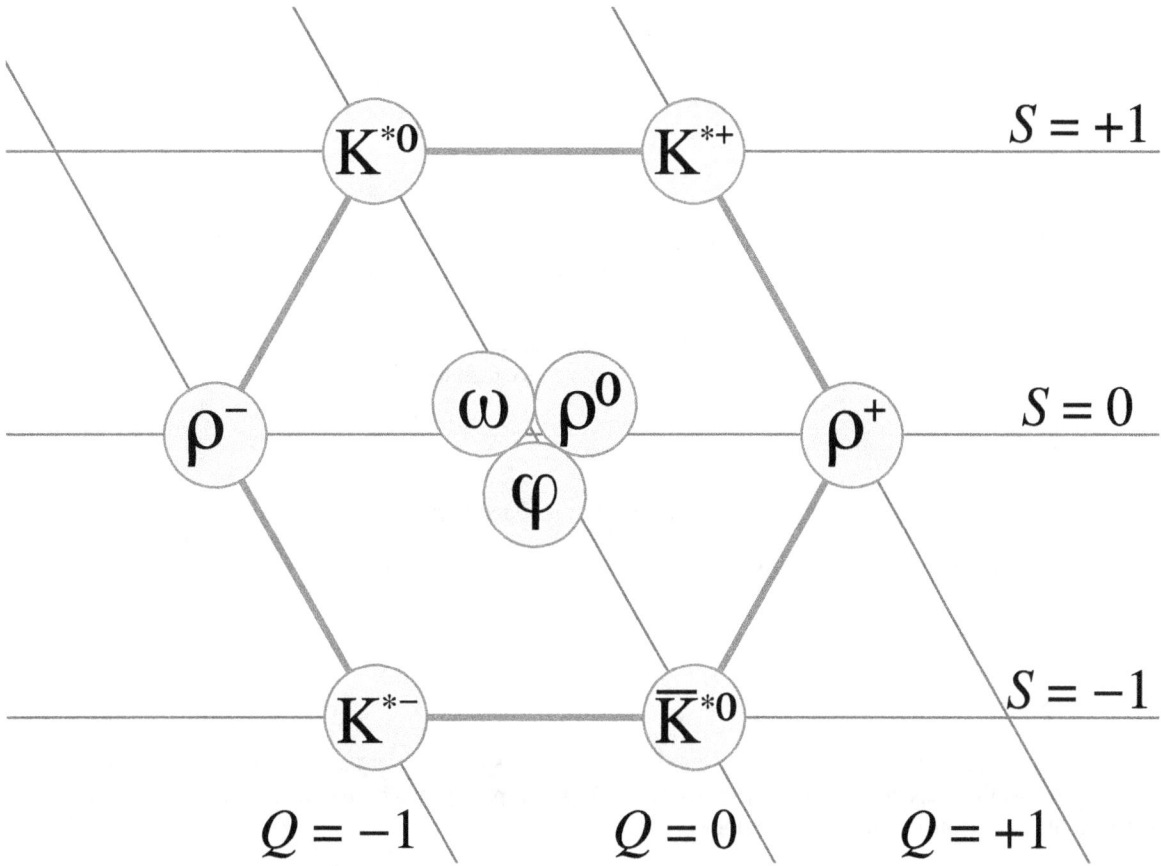

Combinations of one u, d or s quarks and one u, d, or s antiquark in J^P = 1$^-$ configuration also form a nonet.

$$S = -(n_s - n_{\bar{s}})$$
$$C = +(n_c - n_{\bar{c}})$$
$$B' = -(n_b - n_{\bar{b}})$$
$$T = +(n_t - n_{\bar{t}}),$$

meaning that the Gell-Mann–Nishijima formula is equivalent to the expression of charge in terms of quark content:

$$Q = \frac{2}{3}[(n_u - n_{\bar{u}}) + (n_c - n_{\bar{c}}) + (n_t - n_{\bar{t}})] - \frac{1}{3}[(n_d - n_{\bar{d}}) + (n_s - n_{\bar{s}}) + (n_b - n_{\bar{b}})].$$

1.3 Classification

Mesons are classified into groups according to their isospin (*I*), total angular momentum (*J*), parity (*P*), G-parity (*G*) or C-parity (*C*) when applicable, and quark (*q*) content. The rules for classification are defined by the Particle Data Group, and are rather convoluted.[18] The rules are presented below, in table form for simplicity.

1.3.1 Types of meson

Mesons are classified into types according to their spin configurations. Some specific configurations are given special names based on the mathematical properties of their spin configuration.

1.3.2 Nomenclature

Flavourless mesons

Flavourless mesons are mesons made of pair of quark and antiquarks of the same flavour (all their flavour quantum numbers are zero: $S = 0$, $C = 0$, $B' = 0$, $T = 0$).[20] The rules for flavourless mesons are:[18]

† ^ The C parity is only relevant to neutral mesons.
†† ^ For $J^{PC} = 1^{--}$, the ψ is called the J/ψ

In addition:

- When the spectroscopic state of the meson is known, it is added in parentheses.

- When the spectroscopic state is unknown, mass (in MeV/c^2) is added in parentheses.

- When the meson is in its ground state, nothing is added in parentheses.

Flavoured mesons

Flavoured mesons are mesons made of pair of quark and antiquarks of different flavours. The rules are simpler in this case: the main symbol depends on the heavier quark, the superscript depends on the charge, and the subscript (if any) depends on the lighter quark. In table form, they are:[18]

In addition:

- If J^P is in the "normal series" (i.e., $J^P = 0^+$, 1^-, 2^+, 3^-, ...), a superscript $*$ is added.

- If the meson is not pseudoscalar ($J^P = 0^-$) or vector ($J^P = 1^-$), J is added as a subscript.

- When the spectroscopic state of the meson is known, it is added in parentheses.

- When the spectroscopic state is unknown, mass (in MeV/c^2) is added in parentheses.

- When the meson is in its ground state, nothing is added in parentheses.

1.4 Exotic mesons

Main article: Exotic meson

There is experimental evidence for particles that are hadrons (i.e., are composed of quarks) and are color-neutral with zero baryon number, and thus by conventional definition are mesons. Yet, these particles do not consist of a single quark-antiquark pair, as all the other conventional mesons discussed above do. A tentative category for these particles is exotic mesons.

There are at least five exotic meson resonances that have been experimentally confirmed to exist by two or more independent experiments. The most statistically significant of these is the Z(4430), discovered by the Belle experiment in 2007 and confirmed by LHCb in 2014. It is a candidate for being a tetraquark: a particle composed of two quarks and two antiquarks.[21] See the main article above for other particle resonances that are candidates for being exotic mesons.

1.5 List

Main article: List of mesons

1.6 See also

- Standard Model

1.7 Notes

[1] J.J. Aubert *et al.* (1974)

[2] J.E. Augustin *et al.* (1974)

[3] S.W. Herb *et al.* (1977)

[4] The Noble Foundation (1949) Nobel Prize in Physics 1949 – Presentation Speech

[5] H. Yukawa (1935)

[6] G. Gamow (1961)

[7] J. Steinberger (1998)

[8] C. Amsler *et al.* (2008): Quark Model

[9] M.S. Sozzi (2008b)

[10] J.W. Cronin (1980)

[11] V.L. Fitch (1980)

[12] M.S. Sozzi (2008c)

[13] K. Gottfried, V.F. Weisskopf (1986)

[14] W. Heisenberg (1932)

[15] E. Wigner (1937)

[16] M. Gell-Mann (1964)

[17] S.S.M Wong (1998)

[18] C. Amsler *et al.* (2008): Naming scheme for hadrons

[19] W.E. Burcham, M. Jobes (1995)

[20] For the purpose of nomenclature, the isospin projection I_3 isn't considered a flavour quantum number. This means that the charged pion-like mesons ($\pi^\pm$, $a^\pm$, $b^\pm$, and $\rho^\pm$ mesons) follow the rules of flavourless mesons, even if they aren't truly "flavourless".

[21] LHCb collaborators (2014): Observation of the resonant character of the Z(4430)– state

1.8 References

- M.S. Sozzi (2008a). "Parity". *Discrete Symmetries and CP Violation: From Experiment to Theory*. Oxford University Press. pp. 15–87. ISBN 0-19-929666-9.

- M.S. Sozzi (2008b). "Charge Conjugation". *Discrete Symmetries and CP Violation: From Experiment to Theory*. Oxford University Press. pp. 88–120. ISBN 0-19-929666-9.

- M.S. Sozzi (2008c). "CP-Symmetry". *Discrete Symmetries and CP Violation: From Experiment to Theory*. Oxford University Press. pp. 231–275. ISBN 0-19-929666-9.

- C. Amsler *et al.* (Particle Data Group) (2008). "Review of Particle Physics". *Physics Letters B* **667** (1): 1–1340. Bibcode:2008PhLB..667....1P. doi:10.1016/j.physletb.2008.07.018.

- S.S.M. Wong (1998). "Nucleon Structure". *Introductory Nuclear Physics* (2nd ed.). New York (NY): John Wiley & Sons. pp. 21–56. ISBN 0-471-23973-9.

- W.E. Burcham, M. Jobes (1995). *Nuclear and Particle Physics* (2nd ed.). Longman Publishing. ISBN 0-582-45088-8.

- R. Shankar (1994). *Principles of Quantum Mechanics* (2nd ed.). New York (NY): Plenum Press. ISBN 0-306-44790-8.

- J. Steinberger (1989). "Experiments with high-energy neutrino beams". *Reviews of Modern Physics* **61** (3): 533–545. Bibcode:1989RvMP...61..533S. doi:10.1103/RevModPhys.61.533.

- K. Gottfried, V.F. Weisskopf (1986). "Hadronic Spectroscopy: G-parity". *Concepts of Particle Physics* **2**. Oxford University Press. pp. 303–311. ISBN 0-19-503393-0.

- J.W. Cronin (1980). "CP Symmetry Violation—The Search for its origin" (PDF). The Nobel Foundation.

- V.L. Fitch (1980). "The Discovery of Charge—Conjugation Parity Asymmetry" (PDF). The Nobel Foundation.

- S.W. Herb; Hom, D.; Lederman, L.; Sens, J.; Snyder, H.; Yoh, J.; Appel, J.; Brown, B.; et al. (1977). "Observation of a Dimuon Resonance at 9.5 Gev in 400-GeV Proton-Nucleus Collisions". *Physical Review Letters* **39** (5): 252–255. Bibcode:1977PhRvL..39..252H. doi:10.1103/PhysRevLett.39.252.

- J.J. Aubert; Becker, U.; Biggs, P.; Burger, J.; Chen, M.; Everhart, G.; Goldhagen, P.; Leong, J.; et al. (1974). "Experimental Observation of a Heavy Particle J". *Physical Review Letters* **33** (23): 1404–1406. Bibcode:1974PhRvL..33.1404A.doi:10.1103/PhysRevLett.33.1404.

- J.E. Augustin; Boyarski, A.; Breidenbach, M.; Bulos, F.; Dakin, J.; Feldman, G.; Fischer, G.; Fryberger, D.; et al. (1974). "Discovery of a Narrow Resonance in e^+e^- Annihilation". *Physical Review Letters* **33** (23): 1406–1408. Bibcode:1974PhRvL..33.1406A. doi:10.1103/PhysRevLett.33.1406.

- M. Gell-Mann (1964). "A Schematic of Baryons and Mesons".*Physics Letters***8**(3): 214–215.Bibcode:1964PhL doi:10.1016/S0031-9163(64)92001-3.

- Ishfaq Ahmad (1965). "the Interactions of 200 MeV $\pi\pm$ -Mesons with Complex Nuclei Proposal to Study the Interactions of 200 MeV $\pi\pm$ -Mesons with Complex Nuclei" (PDF). *CERN documents* **3** (5).

- G. Gamow (1988) [1961]. *The Great Physicists from Galileo to Einstein* (Reprint ed.). Dover Publications. p. 315. ISBN 978-0-486-25767-9.

- E. Wigner (1937). "On the Consequences of the Symmetry of the Nuclear Hamiltonian on the Spectroscopy of Nuclei". *Physical Review* **51** (2): 106–119. Bibcode:1937PhRv...51..106W. doi:10.1103/PhysRev.51.106.

- H. Yukawa (1935). "On the Interaction of Elementary Particles" (PDF). *Proc. Phys. Math. Soc. Jap.* **17** (48).

- W. Heisenberg (1932). "Über den Bau der Atomkerne I".*Zeitschrift für Physik*(in German)**77**: 1–11.Bibcode:19 doi:10.1007/BF01342433.

- W. Heisenberg (1932). "Über den Bau der Atomkerne II". *Zeitschrift für Physik* (in German) **78** (3–4): 156–164. Bibcode:1932ZPhy...78..156H. doi:10.1007/BF01337585.

- W. Heisenberg (1932). "Über den Bau der Atomkerne III". *Zeitschrift für Physik* (in German) **80** (9–10): 587–596. Bibcode:1933ZPhy...80..587H. doi:10.1007/BF01335696.

1.9   External links

- A table of some mesons and their properties

- *Particle Data Group*—Compiles authoritative information on particle properties

- hep-ph/0211411: The light scalar mesons within quark models

- Naming scheme for hadrons (a PDF file)

- Mesons made thinkable, an interactive visualisation allowing physical properties to be compared

1.9.1   Recent findings

- What Happened to the Antimatter? Fermilab's DZero Experiment Finds Clues in Quick-Change Meson

- CDF experiment's definitive observation of matter-antimatter oscillations in the Bs meson

Chapter 2

Parity (physics)

In quantum physics, a **parity transformation** (also called **parity inversion**) is the flip in the sign of *one* spatial coordinate. In three dimensions, it is also often described by the simultaneous flip in the sign of all three spatial coordinates (a point reflection):

$$\mathbf{P} : \begin{pmatrix} x \\ y \\ z \end{pmatrix} \mapsto \begin{pmatrix} -x \\ -y \\ -z \end{pmatrix}.$$

It can also be thought of as a test for chirality of a physical phenomenon, in that a parity inversion transforms a phenomenon into its mirror image. A parity transformation on something achiral, on the other hand, can be viewed as an identity transformation. All fundamental interactions of elementary particles, with the exception of the weak interaction, are symmetric under parity. The weak interaction is chiral and thus provides a means for probing chirality in physics. In interactions that are symmetric under parity, such as electromagnetism in atomic and molecular physics, parity serves as a powerful controlling principle underlying quantum transitions.

A matrix representation of **P** (in any number of dimensions) has determinant equal to −1, and hence is distinct from a rotation, which has a determinant equal to 1. In a two-dimensional plane, a simultaneous flip of all coordinates in sign is *not* a parity transformation; it is the same as a 180°-rotation.

2.1   Simple symmetry relations

Under rotations, classical geometrical objects can be classified into scalars, vectors, and tensors of higher rank. In classical physics, physical configurations need to transform under representations of every symmetry group.

Quantum theory predicts that states in a Hilbert space do not need to transform under representations of the group of rotations, but only under projective representations. The word *projective* refers to the fact that if one projects out the phase of each state, where we recall that the overall phase of a quantum state is not an observable, then a projective representation reduces to an ordinary representation. All representations are also projective representations, but the converse is not true, therefore the projective representation condition on quantum states is weaker than the representation condition on classical states.

The projective representations of any group are isomorphic to the ordinary representations of a central extension of the group. For example, projective representations of the 3-dimensional rotation group, which is the special orthogonal group SO(3), are ordinary representations of the special unitary group SU(2) (see Representation theory of SU(2)). Projective representations of the rotation group that are not representations are called spinors, and so quantum states may transform not only as tensors but also as spinors.

If one adds to this a classification by parity, these can be extended, for example, into notions of

- *scalars* ($P = 1$) and *pseudoscalars* ($P = -1$) which are rotationally invariant.

- *vectors* ($P = -1$) and *axial vectors* (also called *pseudovectors*) ($P = 1$) which both transform as vectors under rotation.

One can define **reflections** such as

$$V_x : \begin{pmatrix} x \\ y \\ z \end{pmatrix} \mapsto \begin{pmatrix} -x \\ y \\ z \end{pmatrix},$$

which also have negative determinant and form a valid parity transformation. Then, combining them with rotations (or successively performing x-, y-, and z-reflections) one can recover the particular parity transformation defined earlier. The first parity transformation given does not work in an even number of dimensions, though, because it results in a positive determinant. In odd number of dimensions only the latter example of a parity transformation (or any reflection of an odd number of coordinates) can be used.

Parity forms the abelian group Z_2 due to the relation $\mathbf{P}^2 = 1$. All Abelian groups have only one-dimensional irreducible representations. For Z_2, there are two irreducible representations: one is even under parity ($\mathbf{P}\varphi = \varphi$), the other is odd ($\mathbf{P}\varphi = -\varphi$). These are useful in quantum mechanics. However, as is elaborated below, in quantum mechanics states need not transform under actual representations of parity but only under projective representations and so in principle a parity transformation may rotate a state by any phase.

2.2   Classical mechanics

Newton's equation of motion $\mathbf{F} = m\mathbf{a}$ (if the mass is constant) equates two vectors, and hence is invariant under parity. The law of gravity also involves only vectors and is also, therefore, invariant under parity.

However, angular momentum $\mathbf{L}$ is an axial vector,

$$\mathbf{L} = \mathbf{r} \times \mathbf{p},$$
$$\mathbf{P}(\mathbf{L}) = (-\mathbf{r}) \times (-\mathbf{p}) = \mathbf{L}.$$

In classical electrodynamics, the charge density ϱ is a scalar, the electric field, $\mathbf{E}$, and current $\mathbf{j}$ are vectors, but the magnetic field, $\mathbf{H}$ is an axial vector. However, Maxwell's equations are invariant under parity because the curl of an axial vector is a vector.

2.3   Effect of spatial inversion on some variables of classical physics

2.3.1   Even

Classical variables, predominantly scalar quantities, which do not change upon spatial inversion include:

t , the time when an event occurs

m , the mass of a particle

E , the energy of the particle

P , power (rate of work done)

ρ , the electric charge density

V , the electric potential (voltage)

ρ , energy density of the electromagnetic field

L , the angular momentum of a particle (both orbital and spin) (axial vector)

B , the magnetic field (axial vector)

H , the auxiliary magnetic field

M , the magnetization

T_{ij} Maxwell stress tensor.

All masses, charges, coupling constants, and other physical constants, except those associated with the weak force

2.3.2 Odd

Classical variables, predominantly vector quantities, which have their sign flipped by spatial inversion include:

h , the helicity

Φ , the magnetic flux

x , the position of a particle in three-space

v , the velocity of a particle

a , the acceleration of the particle

p , the linear momentum of a particle

F , the force exerted on a particle

J , the electric current density

E , the electric field

D , the electric displacement field

P , the electric polarization

A , the electromagnetic vector potential

S , Poynting vector.

2.4 Quantum mechanics

2.4.1 Possible eigenvalues

In quantum mechanics, spacetime transformations act on quantum states. The parity transformation, **P**, is a unitary operator, in general acting on a state ψ as follows: $\mathbf{P}\psi(r) = e^{i\varphi/2}\psi(-r)$.

One must then have $\mathbf{P}^2\psi(r) = e^{i\varphi}\psi(r)$, since an overall phase is unobservable. The operator $\mathbf{P}^2$, which reverses the parity of a state twice, leaves the spacetime invariant, and so is an internal symmetry which rotates its eigenstates by phases $e^{i\varphi}$. If $\mathbf{P}^2$ is an element e^{iQ} of a continuous U(1) symmetry group of phase rotations, then $e^{-iQ/2}$ is part of this U(1) and so is also a symmetry. In particular, we can define $\mathbf{P}' = \mathbf{P}e^{-iQ/2}$, which is also a symmetry, and so we can choose to call $\mathbf{P}'$ our parity operator, instead of **P**. Note that $\mathbf{P}'^2 = 1$ and so $\mathbf{P}'$ has eigenvalues ± 1. However, when no such symmetry group exists, it may be that all parity transformations have some eigenvalues which are phases other than ± 1.

For electronic wavefunctions, even states are usually indicated by a subscript g for *gerade* (German: even) and odd states by a subscript u for *ungerade* (German: odd). For example, the lowest energy level of the hydrogen molecule ion (H_2^+) is labelled $1\sigma_g$ and the next-lowest $1\sigma_u$.[1]

2.4.2   Consequences of parity symmetry

When parity generates the Abelian group $\mathbb{Z}_2$, one can always take linear combinations of quantum states such that they are either even or odd under parity (see the figure). Thus the parity of such states is ± 1. The parity of a multiparticle state is the product of the parities of each state; in other words parity is a multiplicative quantum number

In quantum mechanics, Hamiltonians are invariant (symmetric) under a parity transformation if $\mathbf{P}$ commutes with the Hamiltonian. In non-relativistic quantum mechanics, this happens for any potential which is scalar, i.e., $V = V(r)$, hence the potential is spherically symmetric. The following facts can be easily proven:

- If $|A\rangle$ and $|B\rangle$ have the same parity, then $\langle A| \mathbf{X} |B\rangle = 0$ where $\mathbf{X}$ is the position operator.

- For a state $|L, L_z\rangle$ of orbital angular momentum $\mathbf{L}$ with z-axis projection L_z, $\mathbf{P}|\mathbf{L}, L_z\rangle = (-1)^L |\mathbf{L}, L_z\rangle$.

- If $[\mathbf{H}, \mathbf{P}] = 0$, then atomic dipole transitions only occur between states of opposite parity.[2]

- If $[\mathbf{H}, \mathbf{P}] = 0$, then a non-degenerate eigenstate of $\mathbf{H}$ is also an eigenstate of the parity operator; i.e., a non-degenerate eigenfunction of $\mathbf{H}$ is either invariant to $\mathbf{P}$ or is changed in sign by $\mathbf{P}$.

Some of the non-degenerate eigenfunctions of $\mathbf{H}$ are unaffected (invariant) by parity $\mathbf{P}$ and the others will be merely reversed in sign when the Hamiltonian operator and the parity operator commute:

$$\mathbf{P} \, \Psi = c \, \Psi,$$

where c is a constant, the eigenvalue of $\mathbf{P}$,

$$\mathbf{P}^2 \Psi = c \mathbf{P} \, \Psi.$$

2.5   Quantum field theory

The intrinsic parity assignments in this section are true for relativistic quantum mechanics as well as quantum field theory.

If we can show that the vacuum state is invariant under parity ($\mathbf{P}|0\rangle = |0\rangle$), the Hamiltonian is parity invariant ($[\mathbf{H}, \mathbf{P}] = 0$) and the quantization conditions remain unchanged under parity, then it follows that every state has good parity, and this parity is conserved in any reaction.

To show that quantum electrodynamics is invariant under parity, we have to prove that the action is invariant and the quantization is also invariant. For simplicity we will assume that canonical quantization is used; the vacuum state is then invariant under parity by construction. The invariance of the action follows from the classical invariance of Maxwell's equations. The invariance of the canonical quantization procedure can be worked out, and turns out to depend on the transformation of the annihilation operator:

$$\mathbf{P}a(\mathbf{p}, \pm)\mathbf{P}^+ = -a(-\mathbf{p}, \pm)$$

where $\mathbf{p}$ denotes the momentum of a photon and $\pm$ refers to its polarization state. This is equivalent to the statement that the photon has odd intrinsic parity. Similarly all vector bosons can be shown to have odd intrinsic parity, and all axial-vectors to have even intrinsic parity.

There is a straightforward extension of these arguments to scalar field theories which shows that scalars have even parity, since

$$\mathbf{P}a(\mathbf{p})\mathbf{P}^+ = a(-\mathbf{p}).$$

This is true even for a complex scalar field. (*Details of spinors are dealt with in the article on the* Dirac equation, *where it is shown that fermions and antifermions have opposite intrinsic parity.*)

With fermions, there is a slight complication because there is more than one spin group.

2.6 Parity in the standard model

2.6.1 Fixing the global symmetries

See also: $(-1)^{\mathrm{F}}$

In the Standard Model of fundamental interactions there are precisely three global internal U(1) symmetry groups available, with charges equal to the baryon number B, the lepton number L and the electric charge Q. The product of the parity operator with any combination of these rotations is another parity operator. It is conventional to choose one specific combination of these rotations to define a standard parity operator, and other parity operators are related to the standard one by internal rotations. One way to fix a standard parity operator is to assign the parities of three particles with linearly independent charges B, L and Q. In general one assigns the parity of the most common massive particles, the proton, the neutron and the electron, to be +1.

Steven Weinberg has shown that if $\mathbf{P}^2 = (-1)^F$, where F is the fermion number operator, then, since the fermion number is the sum of the lepton number plus the baryon number, $F = B + L$, for all particles in the Standard Model and since lepton number and baryon number are charges Q of continuous symmetries e^{iQ}, it is possible to redefine the parity operator so that $\mathbf{P}^2 = 1$. However, if there exist Majorana neutrinos, which experimentalists today believe is quite possible, their fermion number is equal to one because they are neutrinos while their baryon and lepton numbers are zero because they are Majorana, and so $(-1)^F$ would not be embedded in a continuous symmetry group. Thus Majorana neutrinos would have parity $\pm i$.

2.6.2 Parity of the pion

In 1954, a paper by William Chinowsky and Jack Steinberger demonstrated that the pion has negative parity.[3] They studied the decay of an "atom" made from a deuteron (2
1H+) and a negatively charged pion ($\pi-$) in a state with zero orbital angular momentum $L = 0$ into two neutrons (n).

Neutrons are fermions and so obey Fermi–Dirac statistics, which implies that the final state is antisymmetric. Using the fact that the deuteron has spin one and the pion spin zero together with the antisymmetry of the final state they concluded that the two neutrons must have orbital angular momentum $L = 1$. The total parity is the product of the intrinsic parities of the particles and the extrinsic parity of the spherical harmonic function $(-1)^L$. Since the orbital momentum changes from zero to one in this process, if the process is to conserve the total parity then the products of the intrinsic parities of the initial and final particles must have opposite sign. A deuteron nucleus is made from a proton and a neutron, and so using the aforementioned convention that protons and neutrons have intrinsic parities equal to +1 they argued that the parity of the pion is equal to minus the product of the parities of the two neutrons divided by that of the proton and neutron in the deuteron, $(-1)(1)^2/(1)^2$, which is equal to minus one. Thus they concluded that the pion is a pseudoscalar particle.

2.6.3 Parity violation

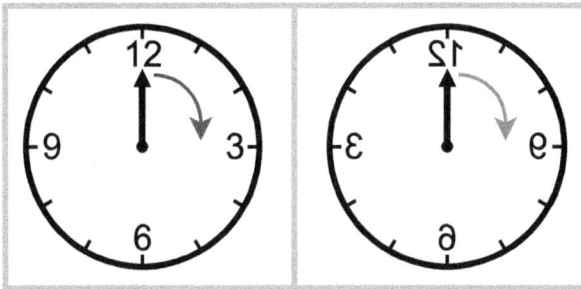

Top: P-symmetry: A clock built like its mirrored image will behave like the mirrored image of the original clock.
Bottom: P-asymmetry: A clock built like its mirrored image will *not* behave like the mirrored image of the original clock.

Although parity is conserved in electromagnetism, strong interactions and gravity, it turns out to be violated in weak interactions. The Standard Model incorporates **parity violation** by expressing the weak interaction as a chiral gauge interaction. Only the left-handed components of particles and right-handed components of antiparticles participate in weak interactions in the Standard Model. This implies that parity is not a symmetry of our universe, unless a hidden mirror sector exists in which parity is violated in the opposite way.

By the mid-20th Century, it had been suggested by several scientists that parity might not be conserved (in different contexts), but without solid evidence these suggestions were not considered important. Then, in 1956, a careful review and analysis by theoretical physicists Tsung Dao Lee and Chen Ning Yang[4] went further, showing that while parity conservation had been verified in decays by the strong or electromagnetic interactions, it was untested in the weak interaction. They proposed several possible direct experimental tests. They were mostly ignored, but Lee was able to convince his Columbia colleague Chien-Shiung Wu to try it. She needed special cryogenic facilities and expertise, so the experiment was done at the National Bureau of Standards.

In 1957 C. S. Wu, E. Ambler, R. W. Hayward, D. D. Hoppes, and R. P. Hudson found a clear violation of parity conservation in the beta decay of cobalt-60.[5] As the experiment was winding down, with double-checking in progress, Wu informed Lee and Yang of their positive results, and saying the results need further examination, she asked them not to publicize the results first. However, Lee revealed the results to his Columbia colleagues on 4 January 1957 at a "Friday Lunch" gathering of the Physics Department of Columbia. Three of them, R. L. Garwin, Leon Lederman, and R. Weinrich modified an existing cyclotron experiment, and they immediately verified the parity violation.[6] They delayed publication of their results until after Wu's group was ready, and the two papers appeared back to back in the same physics journal.

After the fact, it was noted that an obscure 1928 experiment had in effect reported parity violation in weak decays, but since the appropriate concepts had not yet been developed, those results had no impact.[7] The discovery of parity violation immediately explained the outstanding τ–θ puzzle in the physics of kaons.

In 2010, it was reported that physicists working with the Relativistic Heavy Ion Collider (RHIC) had created a short-lived parity symmetry-breaking bubble in quark-gluon plasmas. An experiment conducted by several physicists including Yale's Jack Sandweiss as part of the STAR collaboration, suggested that parity may also be violated in the strong interaction.[8]

2.6.4   Intrinsic parity of hadrons

To every particle one can assign an **intrinsic parity** as long as nature preserves parity. Although weak interactions do not, one can still assign a parity to any hadron by examining the strong interaction reaction that produces it, or through decays not involving the weak interaction, such as rho meson decay to pions.

2.7   See also

- Electroweak theory

- Standard Model

- Mirror matter

2.8 References

General

- Perkins, Donald H. (2000). *Introduction to High Energy Physics*. ISBN 9780521621960.

- Sozzi, M. S. (2008). *Discrete symmetries and CP violation*. Oxford University Press. ISBN 978-0-19-929666-8.

- Bigi, I. I.; Sanda, A. I. (2000). *CP Violation*. Cambridge Monographs on Particle Physics, Nuclear Physics and Cosmology. Cambridge University Press. ISBN 0-521-44349-0.

- Weinberg, S. (1995). *The Quantum Theory of Fields*. Cambridge University Press. ISBN 0-521-67053-5.

Specific

[1] Levine, I.N. *Quantum Chemistry* (Prentice-Hall, 4th edn. 1991), p.355

[2] Bransden, B. H.; Joachain, C. J. (2003). *Physics of Atoms and Molecules* (2nd ed.). Prentice Hall. p. 204. ISBN 978-0-582-35692-4.

[3] Chinowsky, W.; Steinberger, J. (1954). "Absorption of Negative Pions in Deuterium: Parity of the Pion". *Physical Review* **95** (6): 1561–1564. Bibcode:1954PhRv...95.1561C. doi:10.1103/PhysRev.95.1561.

[4] Lee, T. D.; Yang, C. N. (1956). "Question of Parity Conservation in Weak Interactions". *Physical Review* **104** (1): 254–258. Bibcode:1956PhRv..104..254L. doi:10.1103/PhysRev.104.254.

[5] Wu, C. S.; Ambler, E; Hayward, R. W.; Hoppes, D. D.; Hudson, R. P. (1957). "Experimental Test of Parity Conservation in Beta Decay". *Physical Review* **105** (4): 1413–1415. Bibcode:1957PhRv..105.1413W. doi:10.1103/PhysRev.105.1413.

[6] Garwin, R. L.; Lederman, L. M.; Weinrich, M. (1957). "Observations of the Failure of Conservation of Parity and Charge Conjugation in Meson Decays: The Magnetic Moment of the Free Muon". *Physical Review* **105** (4): 1415–1417. Bibcode:1957PhRv..105.1415G.doi:10.1103/PhysRev.105.1415.

[7] Roy, A. (2005). "Discovery of parity violation". *Resonance* **10** (12): 164–175. doi:10.1007/BF02835140.

[8] Muzzin, S. T. (19 March 2010). "For One Tiny Instant, Physicists May Have Broken a Law of Nature". *PhysOrg*. Retrieved 2011-08-05.

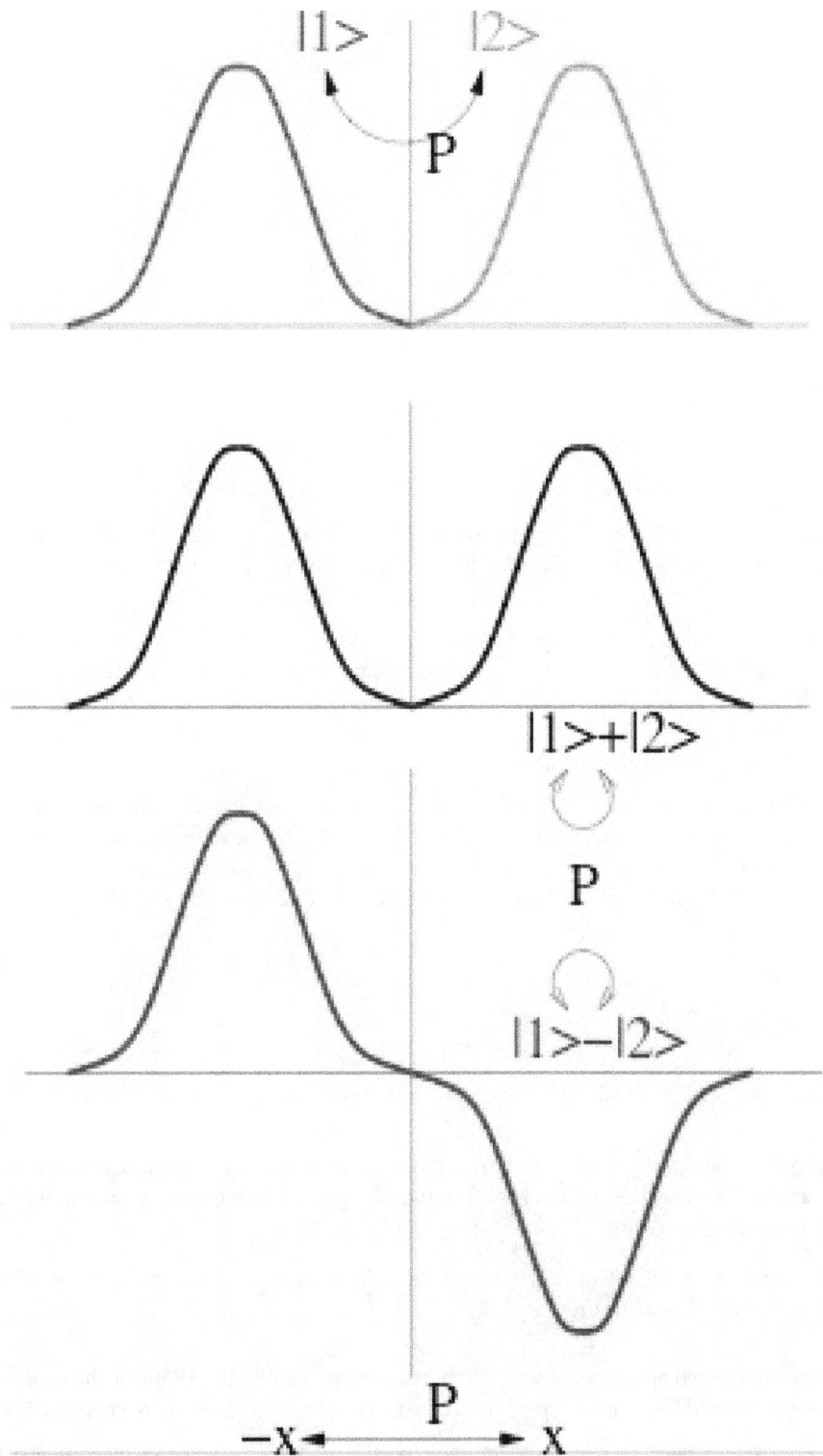

Two dimensional representations of parity are given by a pair of quantum states which go into each other under parity. However, this representation can always be reduced to linear combinations of states, each of which is either even or odd under parity. One says that all irreducible representations of parity are one-dimensional.

Chapter 3

Isospin

In nuclear physics and particle physics, **isospin** (*isotopic spin*, *isobaric spin*) is a quantum number related to the strong interaction. Particles that are affected equally by the strong force but have different charges (e.g. protons and neutrons) can be treated as being different states of the same particle with isospin values related to the number of charge states.[1]

Although it does not have the units of angular momentum and is not a type of spin, the formalism that describes it is mathematically similar to that of angular momentum in quantum mechanics, which means it can be coupled in the same manner. For example, a proton-neutron pair can be coupled in a state of total isospin 1 or 0.[2] It is a dimensionless quantity and the name derives from the fact that the mathematical structures used to describe it are very similar to those used to describe the intrinsic angular momentum (spin).

This term was derived from *isotopic spin*, a confusing term to which nuclear physicists prefer *isobaric spin*, which is more precise in meaning. Isospin symmetry is a subset of the flavour symmetry seen more broadly in the interactions of baryons and mesons. Isospin symmetry remains an important concept in particle physics, and a close examination of this symmetry historically led directly to the discovery and understanding of quarks and of the development of Yang–Mills theory.

3.1   Motivation for isospin

Isospin was introduced by Werner Heisenberg in 1932[3] to explain symmetries of the then newly discovered neutron:

- The mass of the neutron and the proton are almost identical: they are nearly degenerate, and both are thus often called nucleons. Although the proton has a positive charge, and the neutron is neutral, they are almost identical in all other respects.

- The strength of the strong interaction between any pair of nucleons is the same, independent of whether they are interacting as protons or as neutrons.

Thus, isospin was introduced as a concept well before the development in the 1960s of the quark model which provides our modern understanding. The name *isospin* however, was introduced by Eugene Wigner in 1937.[4]

Protons and neutrons, baryons of spin $1/2$, were grouped together as nucleons because they both have nearly the same mass and interact in nearly the same way. Thus, it was convenient to treat them as being different states of the same particle. Since a spin $1/2$ particle has two states, the two were said to be of isospin $1/2$. The proton and neutron were then associated with different isospin projections $I_3 = +1/2$ and $-1/2$ respectively. When constructing a physical theory of nuclear forces, one could then simply assume that it does not depend on isospin.

These considerations would also prove useful in the analysis of meson-nucleon interactions after the discovery of the pions in 1947. The three pions ($\pi+$, $\pi0$, $\pi-$) could be assigned to an isospin triplet with $I = 1$ and $I_3 = +1$, 0 or -1. By assuming that isospin was conserved by nuclear interactions, the new mesons were more easily accommodated by nuclear theory.

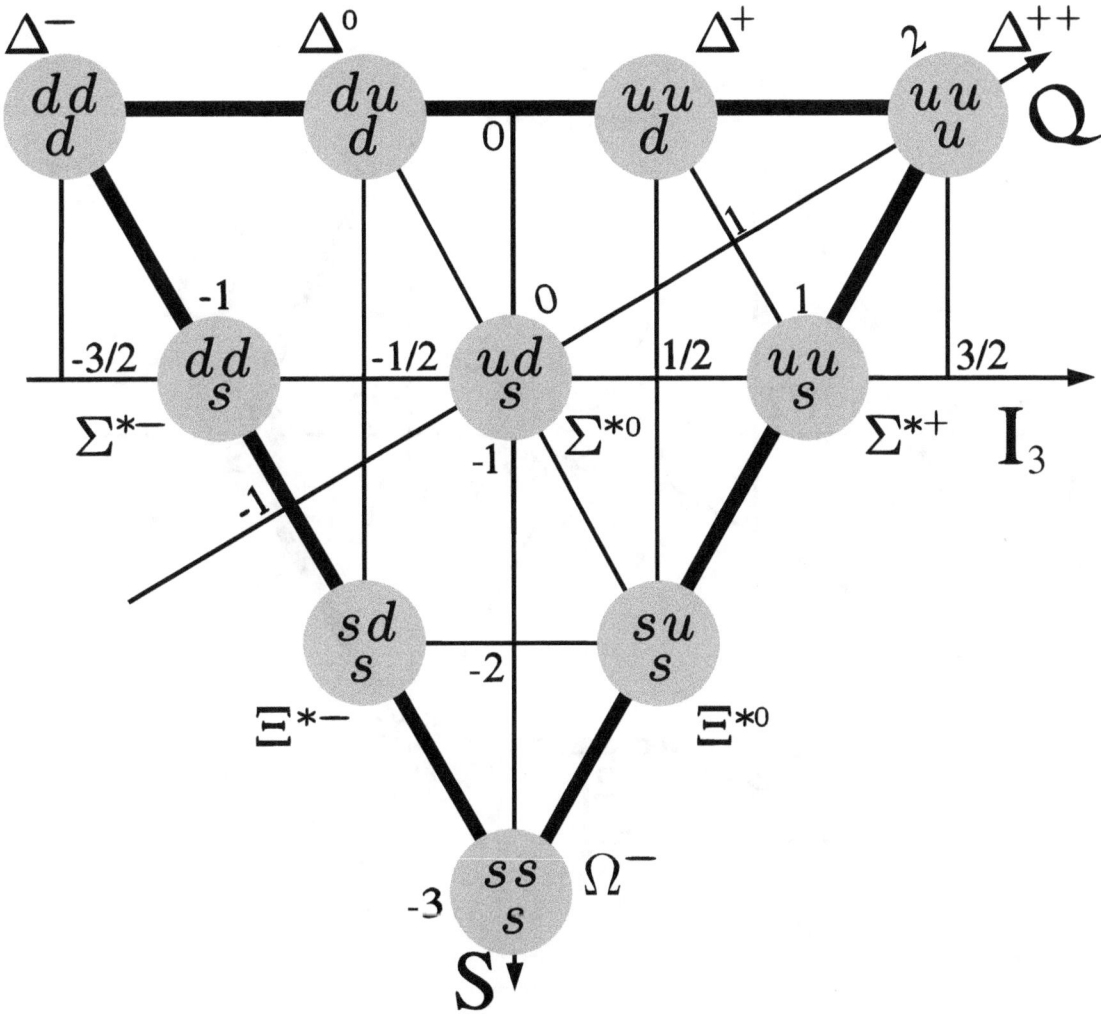

Combinations of three u, d or s-quarks forming baryons with spin-$\frac{3}{2}$ form the baryon decuplet.

As further particles were discovered, they were assigned into isospin multiplets according to the number of different charge states seen: 2 doublets, $I = -\frac{1}{2}$ and $I = \frac{1}{2}$ of K mesons (K−, K0),(K+, K0), a triplet $I = 1$ of Sigma baryons (Σ+, Σ0, Σ−) a singlet $I = 0$ Lambda baryon (Λ0), a quartet $I = \frac{3}{2}$ Delta baryons (Δ++, Δ+, Δ0, Δ−), and so on. This multiplet structure was combined with strangeness in Murray Gell-Mann's eightfold way, ultimately leading to the quark model and quantum chromodynamics.

3.2 Modern understanding of isospin

Observation of the light baryons (those made of up, down and strange quarks) lead us to believe that some of these particles are so similar in terms of their strong interactions that they can be treated as different states of the same particle. In the modern understanding of quantum chromodynamics, this is because up and down quarks are very similar in mass, and have the same strong interactions. Particles made of the same numbers of up and down quarks have similar masses and are grouped together. For examples, the particles known as the Delta baryons—baryons of spin $\frac{3}{2}$ made of a mix of three up and down quarks—are grouped together because they all have nearly the same mass (approximately 1232 MeV/c^2), and interact in nearly the same way.

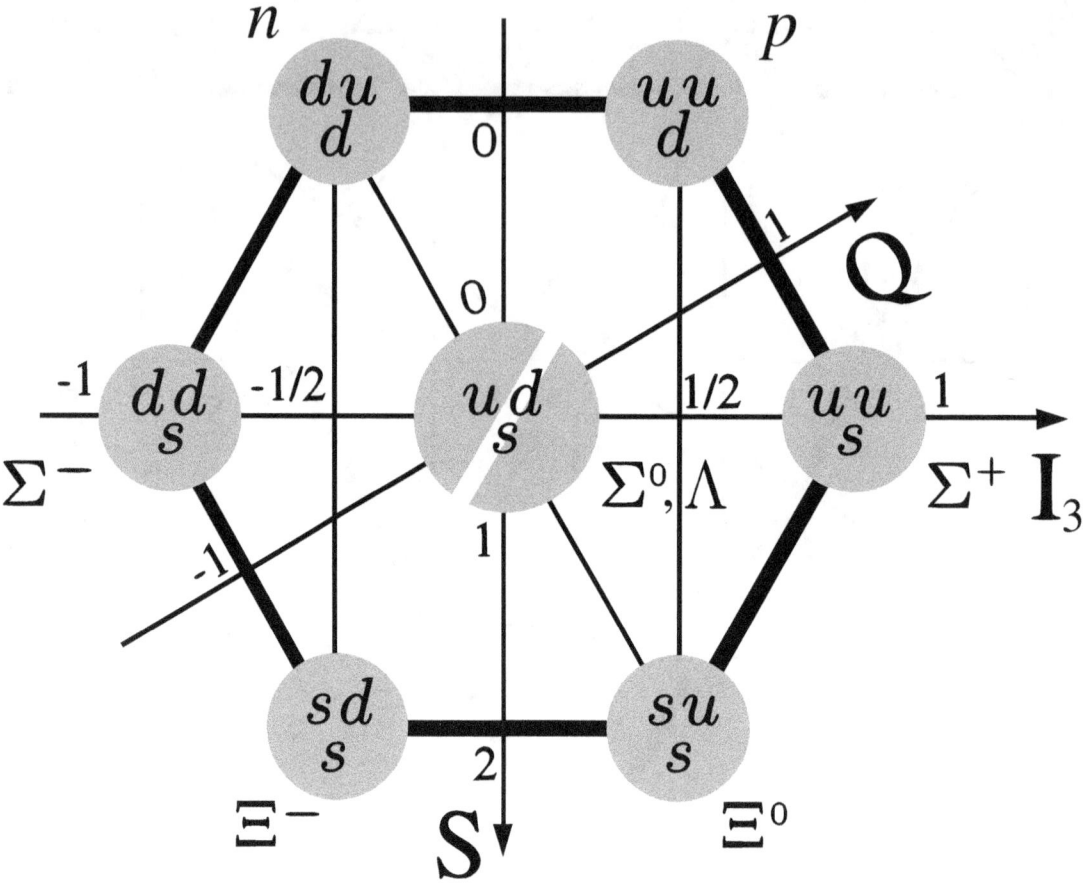

Combinations of three u, d or s-quarks forming baryons with spin-$\frac{1}{2}$ form the baryon octet

However, because the up and down quarks have different charges ($\frac{2}{3}$ e and $-\frac{1}{3}$ e respectively), the four Deltas also have different charges (Δ++ (uuu), Δ+ (uud), Δ0 (udd), Δ- (ddd)). These Deltas could be treated as the same particle and the difference in charge being due to the particle being in different states. Isospin was devised as a parallel to spin to associate an isospin projection (denoted I_3) to each charged state. Since there were four Deltas, four projections were needed. Because isospin was modeled on spin, the isospin projections were made to vary in increments of 1 and to have four increments of 1, you needed an isospin value of $\frac{3}{2}$ (giving the projections $I_3 = \frac{3}{2}$, $\frac{1}{2}$, $-\frac{1}{2}$, $-\frac{3}{2}$). Thus, all the Deltas were said to have isospin $I = \frac{3}{2}$ and each individual charge had different I_3 (e.g. the Δ++ was associated with $I_3 = +\frac{3}{2}$). In the isospin picture, the four Deltas and the two nucleons were thought to be the different states of two particles. In the quark model, the Deltas can be thought of as the excited states of the nucleons.

After the quark model was elaborated, it was noted that the isospin projection was related to the up and down quark content of particles. The relation is

$$I_3 = \frac{1}{2}\left[\left(n_\mathrm{u} - n_{\bar{\mathrm{u}}}\right) - \left(n_\mathrm{d} - n_{\bar{\mathrm{d}}}\right)\right]$$

where n_u and n_d are the numbers of up and down quarks respectively, and n_u and n_d are the numbers of up and down antiquarks respectively.

By this, the value of I_3 of the nucleons proton (symbol p) and neutron (symbol n) is determined by their quark composition, *uud* for the proton and *udd* for the neutron.

3.3   Isospin symmetry

In quantum mechanics, when a Hamiltonian has a symmetry, that symmetry manifests itself through a set of states that have the same energy; that is, the states are degenerate. In particle physics, the near mass-degeneracy of the neutron and proton points to an approximate symmetry of the Hamiltonian describing the strong interactions. The neutron does have a slightly higher mass due to isospin breaking; this is due to the difference in the masses of the up and down quarks and the effects of the electromagnetic interaction. However, the appearance of an approximate symmetry is still useful, since the small breakings can be described by a perturbation theory, which gives rise to slight differences between the near-degenerate states.

3.3.1   SU(2)

See also: Representation theory of SU(2)

Heisenberg's contribution was to note that the mathematical formulation of this symmetry was in certain respects similar to the mathematical formulation of spin, whence the name "isospin" derives. To be precise, the isospin symmetry is given by the invariance of the Hamiltonian of the strong interactions under the action of the Lie group SU(2). The neutron and the proton are assigned to the doublet (the spin-$\frac{1}{2}$, **2**, or fundamental representation) of SU(2). The pions are assigned to the triplet (the spin-1, **3**, or adjoint representation) of SU(2). Though, there is a difference from the theory of spin: the group action does not preserve flavor.

Like the case for regular spin, the isospin operator **I** is vector-valued: it has three components $\mathbf{I}x$, $\mathbf{I}y$, $\mathbf{I}z$ which are coordinates in the same 3-dimensional vector space where the **3** representation acts. Note that it has nothing to do with the physical space, except similar mathematical formalism. Isospin is described by two quantum numbers: I, the total isospin, and I_3, an eigenvalue of the $\mathbf{I}z$ projection for which flavor states are eigenstates, not an *arbitrary projection* as in the case of spin. In other words, each I_3 state specifies certain flavor state of a multiplet. The third coordinate (z), to which the "3" subscript refers, is chosen due to notational conventions which relate bases in **2** and **3** representation spaces. Namely, for the spin-$\frac{1}{2}$ case, components of **I** are equal to Pauli matrices divided by 2 and $\mathbf{I}z = \frac{1}{2}\,\tau_3$, where

$$\tau_3 = \begin{pmatrix} 1 & 0 \\ 0 & -1 \end{pmatrix}$$

While the forms of these matrices are the isomorphic to those of spin, *these* Pauli matrices only acts within the Hilbert space of isospin, not that of spin, and therefore is common to denote them with $\boldsymbol{\tau}$ rather than $\boldsymbol{\sigma}$ to avoid confusion.

The power of isospin symmetry and related methods such as the Eightfold Way come from the observation that families of particles with similar masses tend to correspond to the invariant subspaces associated with the irreducible representations of the Lie algebra $\mathfrak{su}(2)$. In this context, an invariant subspace is spanned by basis vectors which correspond to particles in a family. Under the action of the Lie algebra $\mathfrak{su}(2)$, which generates rotations in isospin space, elements corresponding to definite particle states or superpositions of states can be rotated into each other, but can never leave the space (since the subspace is in fact invariant). This is reflective of the symmetry present. The fact that unitary matrices will commute with the Hamiltonian means that the physical quantities calculated do not change even under unitary transformation. In the case of isospin, this machinery is used to reflect the fact that the strong force behaves the same under the exchange of the up and down quark (and by extension the exchange of the proton and the neutron).

3.4   Relationship to flavor

The discovery and subsequent analysis of additional particles, both mesons and baryons, made it clear that the concept of isospin symmetry could be broadened to an even larger symmetry group, now called flavor symmetry. Once the kaons and their property of strangeness became better understood, it started to become clear that these, too, seemed to be a part of an enlarged symmetry that contained isospin as a subgroup. The larger symmetry was named the Eightfold Way by Murray Gell-Mann, and was promptly recognized to correspond to the adjoint representation of SU(3). To better understand the origin of this symmetry, Gell-Mann proposed the existence of up, down and strange quarks which would belong to the fundamental representation of the SU(3) flavor symmetry.

Although isospin symmetry is very slightly broken, SU(3) symmetry is more badly broken, due to the much higher mass of the strange quark compared to the up and down. The discovery of charm, bottomness and topness could lead to further expansions up to SU(6) flavour symmetry, but the very large masses of these quarks makes such symmetries almost useless. In modern applications, such as lattice QCD, isospin symmetry is often treated as exact while the heavier quarks must be treated separately.

3.5 Quark content and isospin

Up and down quarks each have isospin $I = \frac{1}{2}$, and isospin 3-components (I_3) of $\frac{1}{2}$ and $-\frac{1}{2}$ respectively. All other quarks have $I = 0$. In general

$$I_3 = \frac{1}{2}(n_u - n_d).$$

3.5.1 Hadron nomenclature

Main articles: Baryon and Mesons

Hadron nomenclature is based on isospin.[5]

- Particles of isospin $\frac{3}{2}$ can only be made by a mix of three u and d quarks (Delta baryons).

- Particles of isospin 1 are made of a mix of two u and d quarks (Pi mesons, Rho mesons, Sigma baryons with one heavier quark, etc.).

- Particles of isospin $\frac{1}{2}$ can be made of a mix of three u and d quarks (nucleons) or from one u or d quark with heavier quarks (K mesons, D mesons, Xi baryons, etc.)

- Particles of isospin 0 can be made of one u and one d quark (Eta mesons, Omega mesons, Lambda baryons, etc.), or from no u or d quarks at all (Omega baryons, Phi mesons, etc.), with heavier quarks in all cases.

3.5.2 Isospin symmetry of quarks

In the framework of the Standard Model, the isospin symmetry of the proton and neutron are reinterpreted as the isospin symmetry of the up and down quarks. Technically, the nucleon doublet states are seen to be linear combinations of products of 3-particle isospin doublet states and spin doublet states. That is, the (spin-up) proton wave function, in terms of quark-flavour eigenstates, is described by

$$|p\uparrow\rangle = \frac{1}{3\sqrt{2}}\left(\ |duu\rangle\ \ |udu\rangle\ \ |uud\rangle\ \right)\begin{pmatrix} 2 & -1 & -1 \\ -1 & 2 & -1 \\ -1 & -1 & 2 \end{pmatrix}\begin{pmatrix} |\downarrow\uparrow\uparrow\rangle \\ |\uparrow\downarrow\uparrow\rangle \\ |\uparrow\uparrow\downarrow\rangle \end{pmatrix} \text{[6]}$$

and the (spin-up) neutron by

$$|n\uparrow\rangle = \frac{1}{3\sqrt{2}}\left(\ |udd\rangle\ \ |dud\rangle\ \ |ddu\rangle\ \right)\begin{pmatrix} 2 & -1 & -1 \\ -1 & 2 & -1 \\ -1 & -1 & 2 \end{pmatrix}\begin{pmatrix} |\downarrow\uparrow\uparrow\rangle \\ |\uparrow\downarrow\uparrow\rangle \\ |\uparrow\uparrow\downarrow\rangle \end{pmatrix} \text{[6]}$$

Here, $|u\rangle$ is the up quark flavour eigenstate, and $|d\rangle$ is the down quark flavour eigenstate, while $|\uparrow\rangle$ and $|\downarrow\rangle$ are the eigenstates of S_z. Although these superpositions are the technically correct way of denoting a proton and neutron in terms of quark flavour and spin eigenstates, for brevity, they are often simply referred to as "*uud*" and "*udd*". Note also that the derivation above assumes exact isospin symmetry and is modified by SU(2)-breaking terms.

Similarly, the isospin symmetry of the pions are given by:

$$|\pi^+\rangle = |u\bar{d}\rangle$$
$$|\pi^0\rangle = \frac{1}{\sqrt{2}}\left(|u\bar{u}\rangle - |d\bar{d}\rangle\right)$$
$$|\pi^-\rangle = -|d\bar{u}\rangle$$

3.5.3 Weak isospin

Main article: weak isospin

Isospin is similar to, but should not be confused with weak isospin. Briefly, weak isospin is the gauge symmetry of the weak interaction which connects quark and lepton doublets of left-handed particles in all generations; for example, up and down quarks, top and bottom quarks, electrons and electron neutrinos. By contrast (strong) isospin connects only up and down quarks, acts on both chiralities (left and right) and is a global (not a gauge) symmetry.

3.6 Gauged isospin symmetry

Attempts have been made to promote isospin from a global to a local symmetry. In 1954, Chen Ning Yang and Robert Mills suggested that the notion of protons and neutrons, which are continuously rotated into each other by isospin, should be allowed to vary from point to point. To describe this, the proton and neutron direction in isospin space must be defined at every point, giving local basis for isospin. A gauge connection would then describe how to transform isospin along a path between two points.

This Yang–Mills theory describes interacting vector bosons, like the photon of electromagnetism. Unlike the photon, the SU(2) gauge theory would contain self-interacting gauge bosons. The condition of gauge invariance suggests that they have zero mass, just as in electromagnetism.

Ignoring the massless problem, as Yang and Mills did, the theory makes a firm prediction: the vector particle should couple to all particles of a given isospin *universally*. The coupling to the nucleon would be the same as the coupling to the kaons. The coupling to the pions would be the same as the self-coupling of the vector bosons to themselves.

When Yang and Mills proposed the theory, there was no candidate vector boson. J. J. Sakurai in 1960 predicted that there should be a massive vector boson which is coupled to isospin, and predicted that it would show universal couplings. The rho mesons were discovered a short time later, and were quickly identified as Sakurai's vector bosons. The couplings of the rho to the nucleons and to each other were verified to be universal, as best as experiment could measure. The fact that the diagonal isospin current contains part of the electromagnetic current led to the prediction of rho-photon mixing and the concept of vector meson dominance, ideas which led to successful theoretical pictures of GeV-scale photon-nucleus scattering.

Although the discovery of the quarks led to reinterpretation of the rho meson as a vector bound state of a quark and an antiquark, it is sometimes still useful to think of it as the gauge boson of a hidden local symmetry[7]

3.7 References

[1] http://www.thefreedictionary.com/isospin

[2] Povh, Bogdan; Klaus, Rith; Scholz, Christoph; Zetsche, Frank (2008) [1993]. "2". *Particles and Nuclei*. p. 21. ISBN 978-3-540-79367-0.

[3]Heisenberg, W.(1932). "Über den Bau der Atomkerne".*Zeitschrift für Physik*(in German)**77**: 1–11.Bibcode:1932ZPhy...77... doi:10.1007/BF01342433.

[4] Wigner, E. (1937). "On the Consequences of the Symmetry of the Nuclear Hamiltonian on the Spectroscopy of Nuclei". *Physical Review* **51** (2): 106–119. Bibcode:1937PhRv...51..106W. doi:10.1103/PhysRev.51.106.

[5] C. Amsler et al.; (Particle Data Group) (2008). "Review of Particle Physics: Naming scheme for hadrons" (PDF). *Physics Letters B* **667**: 1. Bibcode:2008PhLB..667....1P. doi:10.1016/j.physletb.2008.07.018.

[6] Greiner, W.; Müller, B. (1989). *Quantum Mechanics: Symmetries*. Springer-Verlag. p. 279. ISBN 3-540-58080-8.

[7] Bando, M.; Kugo, T.; Uehara, S.; Yamawaki, K.; Yanagida, T. (1985). "Is the ρ Meson a Dynamical Gauge Boson of Hidden Local Symmetry?". *Physical Review Letters* **54** (12): 1215–1218. Bibcode:1985PhRvL..54.1215B. doi:10.1103/PhysRevLett.54.1215.PMID10030967.

3.8 Further reading

- Itzykson, C.; Zuber, J.-B. (1980). *Quantum Field Theory*. McGraw-Hill. ISBN 0-07-032071-3.

- Griffiths, D. (1987). *Introduction to Elementary Particles*. John Wiley & Sons. ISBN 0-471-60386-4.

3.9 External links

- **Nuclear Structure and Decay Data - IAEA** Nuclides' Isospin

Chapter 4

Flavour (particle physics)

In particle physics, **flavour** or **flavor** refers to a species of an elementary particle. The Standard Model counts six flavours of quarks and six flavours of leptons. They are conventionally parameterized with *flavour quantum numbers* that are assigned to all subatomic particles, including composite ones. For hadrons, these quantum numbers depend on the numbers of constituent quarks of each particular flavour.

4.1   Intuitive description

Elementary particles are not eternal and indestructible. Unlike in classical mechanics, where forces only change a particle's momentum, the weak force can alter the essence of a particle, even an elementary particle. This means that it can convert one quark to another quark with different mass and electric charge, and the same for leptons. From the point of view of quantum mechanics, changing the flavour of a particle by the weak force is no different in principle from changing its spin by electromagnetic interaction, and should be described with quantum numbers as well. In particular, flavour states may undergo quantum superposition.

In atomic physics the principal quantum number of an electron specifies the electron shell in which it resides, which determines the energy level of the whole atom. In an analogous way, the five flavour quantum numbers of a quark specify which of six flavours (u, d, s, c, b, t) it has, and when these quarks are combined this results in different types of baryons and mesons with different masses, electric charges, and decay modes.

4.2   Flavour symmetry

If there are two or more particles which have identical interactions, then they may be interchanged without affecting the physics. Any (complex) linear combination of these two particles give the same physics, as long as they are orthogonal or perpendicular to each other. In other words, the theory possesses symmetry transformations such as $M \begin{pmatrix} u \\ d \end{pmatrix}$, where u and d are the two fields, and M is any 2×2 unitary matrix with a unit determinant. Such matrices form a Lie group called SU(2) (see special unitary group). This is an example of flavour symmetry.

In quantum chromodynamics, flavour is a global symmetry. In the electroweak theory, on the other hand, this symmetry is broken, and flavour changing processes exist, such as quark decay or neutrino oscillations.

4.3   Flavour quantum numbers

4.3.1 Leptons

All leptons carry a lepton number $L = 1$. In addition, leptons carry weak isospin, T_3, which is $-1/2$ for the three charged leptons (i.e. electron, muon and tau) and $+1/2$ for the three associated neutrinos. Each doublet of a charged lepton and a neutrino consisting of opposite T_3 are said to constitute one generation of leptons. In addition, one defines a quantum number called weak hypercharge, YW, which is -1 for all left-handed leptons.[1] Weak isospin and weak hypercharge are gauged in the Standard Model.

Leptons may be assigned the six flavour quantum numbers: electron number, muon number, tau number, and corresponding numbers for the neutrinos. These are conserved in strong and electromagnetic interactions, but violated by weak interactions. Therefore, such flavour quantum numbers are not of great use. A separate quantum number for each generation is more useful: electronic lepton number (+1 for electrons and electron neutrinos), muonic lepton number (+1 for muons and muon neutrinos), and tauonic lepton number (+1 for tau leptons and tau neutrinos). However, even these numbers are not absolutely conserved, as neutrinos of different generations can mix; that is, a neutrino of one flavour can transform into another flavour. The strength of such mixings is specified by a matrix called the Pontecorvo–Maki–Nakagawa–Sakata matrix (PMNS matrix).

4.3.2 Quarks

All quarks carry a baryon number $B = 1/3$. They also all carry weak isospin, $T_3 = \pm 1/2$. The positive-T_3 quarks (up, charm, and top quarks) are called *up-type quarks* and negative-T_3 quarks (down, strange, and bottom quarks) are called *down-type quarks*. Each doublet of up and down type quarks constitutes one generation of quarks.

For all the quark flavour quantum numbers (strangeness, charm, topness and bottomness) the convention is that the flavour charge and the electric charge of a quark have the same sign. Thus any flavour carried by a charged meson has the same sign as its charge. Quarks have the following flavour quantum numbers:

- Isospin, less ambiguously known as "isobaric spin", which has value $I_3 = 1/2$ for the up quark and $I_3 = -1/2$ for the down quark.

- Strangeness (S): Defined as $S = -(n_s - \bar{n}_s)$, where n_s represents the number of strange quarks (s) and $\bar{n}_s$ represents the number of strange antiquarks (s). This quantum number was introduced by Murray Gell-Mann. This definition gives the strange quark a strangeness of -1 for the above-mentioned reason.

- Charm (C): Defined as $C = (n_c - \bar{n}_c)$, where n_c represents the number of charm quarks (c) and $\bar{n}_c$ represents the number of charm antiquarks. Is $+1$ for the charm quark.

- Bottomness (B'): Also called 'beauty'. Defined as $B' = -(n_b - \bar{n}_b)$, where n_b represents the number of bottom quarks (b) and $\bar{n}_b$ represents the number of bottom antiquarks.

- Topness (T): Also called 'truth'. Defined as $T = (n_t - \bar{n}_t)$, where n_t represents the number of top quarks (t) and $\bar{n}_t$ represents the number of top antiquarks. However, because of the extremely short half-life of the top quark, by the time it can interact strongly it has already decayed to another flavour of quark (usually to a bottom quark). For that reason the top quark doesn't hadronize, that is it never forms any meson or baryon.

These five quantum numbers, together with baryon number (which is not a flavour quantum number) completely specify numbers of all 6 quark flavours separately (as $n_q - \bar{n}_q$, i.e. an antiquark is counted with the minus sign). They are conserved by both the electromagnetic and strong interactions (but not the weak interaction). From them can be built the derived quantum numbers:

- Hypercharge (Y): $Y = B + S + C + B' + T$

- Electric charge: $Q = I_3 + 1/2Y$ (see Gell-Mann–Nishijima formula)

The terms "strange" and "strangeness" predate the discovery of the quark, but continued to be used after its discovery for the sake of continuity (i.e. the strangeness of each type of hadron remained the same); strangeness of anti-particles being referred to as +1, and particles as −1 as per the original definition. Strangeness was introduced to explain the rate of decay of newly discovered particles, such as the kaon, and was used in the Eightfold Way classification of hadrons and in subsequent quark models. These quantum numbers are preserved under strong and electromagnetic interactions, but not under weak interactions.

For first-order weak decays, that is processes involving only one quark decay, these quantum numbers (e.g. charm) can only vary by 1 ($|C| = \pm 1$); $\Delta B' = \pm 1$. Since first-order processes are more common than second-order processes (involving two quark decays), this can be used as an approximate "selection rule" for weak decays.

A quark of a given flavour is an eigenstate of the weak interaction part of the Hamiltonian: it will interact in a definite way with the W and Z bosons. On the other hand, a fermion of a fixed mass (an eigenstate of the kinetic and strong interaction parts of the Hamiltonian) is normally a superposition of various flavours. As a result, the flavour content of a quantum state may change as it propagates freely. The transformation from flavour to mass basis for quarks is given by the Cabibbo–Kobayashi–Maskawa matrix (CKM matrix). This matrix is analogous to the PMNS matrix for neutrinos, and defines the strength of flavour changes under weak interactions of quarks.

The CKM matrix allows for CP violation if there are at least three generations.

4.3.3   Antiparticles and hadrons

Flavour quantum numbers are additive. Hence antiparticles have flavour equal in magnitude to the particle but opposite in sign. Hadrons inherit their flavour quantum number from their valence quarks: this is the basis of the classification in the quark model. The relations between the hypercharge, electric charge and other flavour quantum numbers hold for hadrons as well as quarks.

4.4   Quantum chromodynamics

Flavour symmetry is closely related to chiral symmetry. This part of the article is best read along with the one on chirality.

Quantum chromodynamics (QCD) contains six flavours of quarks. However, their masses differ and as a result they are not strictly interchangeable with each other. The up and down flavours are close to having equal masses, and the theory of these two quarks possesses an approximate SU(2) symmetry (isospin symmetry).

Under some circumstances, the masses of the quarks can be neglected entirely. One can then make flavour transformations independently on the left- and right-handed parts of each quark field. The flavour group is then a chiral group SUL(N_f) × SUR(N_f).

If all quarks had non-zero but equal masses, then this chiral symmetry is broken to the *vector symmetry* of the "diagonal flavour group" SU(N_f), which applies the same transformation to both helicities of the quarks. Such a reduction of the symmetry is called *explicit symmetry breaking*. The amount of explicit symmetry breaking is controlled by the current quark masses in QCD.

Even if quarks are massless, chiral flavour symmetry can be spontaneously broken if the vacuum of the theory contains a chiral condensate (as it does in low-energy QCD). This gives rise to an effective mass for the quarks, often identified with the valence quark mass in QCD.

4.4.1   Symmetries of QCD

Analysis of experiments indicate that the current quark masses of the lighter flavours of quarks are much smaller than the QCD scale, ΛQCD, hence chiral flavour symmetry is a good approximation to QCD for the up, down and strange quarks. The success of chiral perturbation theory and the even more naive chiral models spring from this fact. The valence quark masses extracted from the quark model are much larger than the current quark mass. This indicates that QCD has spontaneous chiral symmetry breaking with the formation of a chiral condensate. Other phases of QCD may break the chiral flavour symmetries in other ways.

4.5 Conservation laws

All of the various charges discussed above are conserved by the fact that the charge operator is best understood as the generator of a symmetry that commutes with the Hamiltonian. Thus, the eigenvalues of the various charge operators are conserved.

Absolutely conserved flavour quantum numbers are: (including the baryon number for completeness)

- electric charge (Q)

- weak isospin (I_3)

- baryon number (B)

- lepton number (L)

In some theories, the individual baryon and lepton number conservation can be violated, if the difference between them ($B - L$) is conserved (see chiral anomaly). All other flavour quantum numbers are violated by the electroweak interactions. Strong interactions conserve all flavours.

4.6 History

Some of the historical events that lead to the development of flavour symmetry are discussed in the article on isospin.

4.7 See also

- Standard Model (mathematical formulation)

- Cabibbo–Kobayashi–Maskawa matrix

- Strong CP problem and chirality (physics)

- Chiral symmetry breaking and quark matter

- Quark flavour tagging, such as B-tagging, is an example of particle identification in experimental particle physics.

4.8 References

[1] See table in S. Raby, R. Slanky (1997). "Neutrino Masses: How to add them to the Standard Model" (PDF). *Los Alamos Science* (25): 64.

4.9 Further reading

- Lessons in Particle Physics Luis Anchordoqui and Francis Halzen, University of Wisconsin, 18th Dec. 2009

4.10 External links

- The particle data group.

Chapter 5

Strangeness

This article is about a concept in particle physics. For the definition of "strangeness", see wikt:strangeness. For other uses, see Strange (disambiguation).

In particle physics, **strangeness** ("S") is a property of particles, expressed as a quantum number, for describing decay of particles in strong and electromagnetic reactions, which occur in a short period of time. The strangeness of a particle is defined as:

$$S = -(n_s - n_{\bar{s}})$$

where n_s represents the number of strange quarks (s) and n_s represents the number of strange antiquarks (s).

The terms *strange* and *strangeness* predate the discovery of the quark, and were adopted after its discovery in order to preserve the continuity of the phrase; strangeness of anti-particles being referred to as +1, and particles as −1 as per the original definition. For all the quark flavor quantum numbers (strangeness, charm, topness and bottomness) the convention is that the flavor charge and the electric charge of a quark have the same sign. With this, any flavor carried by a charged meson has the same sign as its charge.

5.1 Conservation

Strangeness was introduced by Murray Gell-Mann and Kazuhiko Nishijima to explain the fact that certain particles, such as the kaons or certain hyperons, were created easily in particle collisions, yet decayed much more slowly than expected for their large masses and large production cross sections. Noting that collisions seemed to always produce pairs of these particles, it was postulated that a new conserved quantity, dubbed "strangeness", was preserved during their creation, but *not* conserved in their decay.

In our modern understanding, strangeness is conserved during the strong and the electromagnetic interactions, but not during the weak interactions. Consequently, the lightest particles containing a strange quark cannot decay by the strong interaction, and must instead decay via the much slower weak interaction. In most cases these decays change the value of the strangeness by one unit. However, this doesn't necessarily hold in second-order weak reactions, where there are mixes of K0 and K0 mesons. All in all, the amount of strangeness can change in a weak interaction reaction by +1, 0 or −1 (depending on the reaction).

5.2 See also

- Strangeness production

5.3 References

- D.J. Griffiths (1987). *Introduction to Elementary Particles*. John Wiley & Sons. ISBN 0-471-60386-4.

5.4 Further reading

- Lessons in Particle Physics Luis Anchordoqui and Francis Halzen, University of Wisconsin, 18th Dec. 2009

Chapter 6

Baryon number

In particle physics, the **baryon number** is a strictly conserved additive quantum number of a system. It is defined as

$$B = \frac{1}{3}\left(n_{\mathrm{q}} - n_{\bar{\mathrm{q}}}\right),$$

where n_{q} is the number of quarks, and n_{q} is the number of antiquarks. Baryons (three quarks) have a baryon number of +1, mesons (one quark, one antiquark) have a baryon number of 0, and antibaryons (three antiquarks) have a baryon number of −1. Exotic hadrons like pentaquarks (four quarks, one antiquark) and tetraquarks (two quarks, two antiquarks) are also classified as baryons and mesons depending on their baryon number.

6.1 Baryon number vs. quark number

See also: Color charge

Quarks carry not only electric charge, but also charges such as color charge and weak isospin. Because of a phenomenon known as *color confinement*, a hadron cannot have a net color charge; that is, the total color charge of a particle has to be zero ("white"). A quark can have one of three "colors", dubbed "red", "green", and "blue".

For normal hadrons, a white color can thus be achieved in one of three ways:

- A quark of one color with an antiquark of the corresponding anticolor, giving a meson with baryon number 0,

- Three quarks of different colors, giving a baryon with baryon number +1,

- Three antiquarks into an antibaryon with baryon number −1.

The baryon number was defined long before the quark model was established, so rather than changing the definitions, particle physicists simply gave quarks one third the baryon number. Nowadays it might be more accurate to speak of the conservation of **quark number**.

In theory, exotic hadrons can be formed by adding pairs of quark and antiquark, provided that each pair has a matching color/anticolor. For example, a pentaquark (four quarks, one antiquark) could have the individual quark colors: red, green, blue, blue, and antiblue.

6.2 Particles not formed of quarks

Particles without any quarks have a baryon number of zero. Such particles include leptons (electron, muon, tau and their neutrinos) and gauge bosons (photon, W and Z bosons, gluons, and the Higgs boson); or the hypothetical graviton.

6.3 Conservation

See also: Conservation law (physics)

The baryon number is conserved in nearly all the interactions of the Standard Model. 'Conserved' means that the sum of the baryon number of all incoming particles is the same as the sum of the baryon numbers of all particles resulting from the reaction. An exception is the chiral anomaly proposed by some extensions of the standard model. However, sphalerons are not all that common. Electroweak sphalerons can only change the baryon number by 3. No experimental evidence of sphalerons has yet been observed.

The still hypothetical idea of a grand unified theory allows for the changing of a baryon into several leptons (see $B - L$), thus violating the conservation of both baryon and lepton numbers.[1] Proton decay would be an example of such a process taking place, but has never been observed.

6.4 See also

- Lepton number

- Flavour (particle physics)

- Isospin

- Hypercharge

- Proton decay

- $B - L$

6.5 References

[1] Griffiths, David (2008). *Introduction to Elementary Particles* (2nd ed.). New York: John Wiley & Sons. p. 77. ISBN 9783527618477. In the grand unified theories new interactions are contemplated, permitting decays such as p+ → e+ + π0 or p+ → ν
μ + π+ in which baryon number and lepton number change.

Chapter 7

Charm (quantum number)

Charm (symbol C) is a flavour quantum number representing the difference between the number of charm quarks (c) and charm antiquarks (c) that are present in a particle:

$$C = n_{\text{c}} - n_{\text{c}}.$$

By convention, the sign of flavour quantum numbers agree with the sign of the electric charge carried by the quark of corresponding flavour. The charm quark, which carries an electric charge (Q) of $+\frac{2}{3}$, therefore carries a charm of $+1$. The charm antiquarks have the opposite charge ($Q = -\frac{2}{3}$), and flavour quantum numbers ($C = -1$).

As with any flavour-related quantum numbers, charm is preserved under strong and electromagnetic interaction, but not under weak interaction (see CKM matrix). For first-order weak decays, that is processes involving only one quark decay, charm can only vary by 1 ($\Delta C = \pm 1, 0$). Since first-order processes are more common than second-order processes (involving two quark decays), this can be used as an approximate "selection rule" for weak decays.

7.1 Further reading

- Lessons in Particle Physics Luis Anchordoqui and Francis Halzen, University of Wisconsin, 18th Dec. 2009

Chapter 8

Bottomness

In physics, **bottomness** (symbol B') also called **beauty**, is a flavour quantum number reflecting the difference between the number of bottom antiquarks (n_b) and the number of bottom quarks (n_b) that are present in a particle:

$$B' = -(n_b - n_{\bar{b}})$$

Bottom quarks have (by convention) a bottomness of −1 while bottom antiquarks have a bottomness of +1. The convention is that the flavour quantum number sign for the quark is the same as the sign of the electric charge (symbol Q) of that quark (in this case, $Q = -\frac{1}{3}$).

As with other flavour-related quantum numbers, bottomness is preserved under strong and electromagnetic interactions, but not under weak interactions. For first-order weak reactions, it holds that $\Delta B' = \pm 1$.

This term is rarely used. Most physicists simply refer to "the number of bottom quarks" and "the number of bottom antiquarks".

8.1 Further reading

- Anchordoqui, L.; Halzen, F. (2009). "Lessons in Particle Physics". arXiv:0906.1271 [physics.ed-ph].

Chapter 9

Topness

Topness (also called **truth**), a flavour quantum number, represents the difference between the number of top quarks (t) and number of top antiquarks (t) that are present in a particle:

$$T = n_t - n_{\bar{t}}$$

By convention, top quarks have a topness of +1 and top antiquarks have a topness of −1. The term "topness" is rarely used; most physicists simply refer to "the number of top quarks" and "the number of top antiquarks".

9.1 Conservation

Like all flavour quantum numbers, topness is preserved under strong and electromagnetic interactions, but not under weak interaction. However the top quark is extremely unstable, with a half-life under 10^{-23} s, which is the required time for the strong interaction to take place. For that reason the top quark does not hadronize, that is it never forms any meson or baryon, so the topness of a meson or a baryon is every time equal at zero. By the time it can interact strongly it has already decayed to another flavour of quark (usually to a bottom quark).

9.2 Further reading

- Anchordoqui, L.; Halzen, F. (2009). "Lessons in Particle Physics". arXiv:0906.1271 [physics.ed-ph].

Chapter 10

D meson

The **D mesons** are the lightest particle containing charm quarks. They are often studied to gain knowledge on the weak interaction.[1] The strange D mesons (D_s) were called the "F mesons" prior to 1986.

10.1   Overview

The D mesons were discovered in 1976 by the Mark I detector at the Stanford Linear Accelerator Center.[2]

Since the D mesons are the lightest mesons containing a single charm quark (or antiquark), they must change the charm (anti)quark into an (anti)quark of another type to decay. Such transitions violate the internal charm quantum number, and can take place only via the weak interaction. In D mesons, the charm quark preferentially changes into a strange quark via an exchange of a W particle, therefore the D meson preferentially decays into Ks and πs.[1]

In November 2011, researchers at the LHCb experiment at CERN reported (3.5 sigma significance) that they have observed a direct CP violation in the neutral D meson decay, possibly beyond the Standard Model.[3]

10.2   List of D mesons

[a] ^ PDG reports the resonance width (Γ). Here the conversion $\tau = \hbar/\Gamma$ is given instead.

10.3   See also

- List of mesons
- List of baryons
- List of particles
- Timeline of particle discoveries

10.4   References

[1] D Meson

[2] http://www.kudryavtsev.staff.shef.ac.uk/phy466/charmed-mesons_files/charmed-mesons.ppt

[3] New Physics at LHC? An Anomaly in CP Violation : Cosmic Variance

[4] C. Amsler *et al.*. (2008): Quark Model

[5] C. Amsler *et al.*. (2008): Particle listings – D±

[6] C. Amsler *et al.*. (2008): Particle listings – D0

[7] N. Nakamura *et al.* (2010): Particle listings – D±
     s

[8] C. Amsler *et al.*. (2008): Particle listings – D∗±(2010)

[9] C. Amsler *et al.*. (2008): Particle listings – D∗0(2007)

Chapter 11

Kaon

For other uses, see Kaon (disambiguation).

In particle physics, a **kaon** /ˈkeɪ.ɒn/, also called a **K meson** and denoted K,[nb 1] is any of a group of four mesons distinguished by a quantum number called strangeness. In the quark model they are understood to be bound states of a strange quark (or antiquark) and an up or down antiquark (or quark).

Kaons have proved to be a copious source of information on the nature of fundamental interactions since their discovery in cosmic rays in 1947. They were essential in establishing the foundations of the Standard Model of particle physics, such as the quark model of hadrons and the theory of quark mixing (the latter was acknowledged by a Nobel Prize in Physics in 2008). Kaons have played a distinguished role in our understanding of fundamental conservation laws: CP violation, a phenomenon generating the observed matter–antimatter asymmetry of the universe, was discovered in the kaon system in 1964 (which was acknowledged by a Nobel Prize in 1980). Moreover, direct CP violation was also discovered in the kaon decays in the early 2000s.

11.1   Basic properties

The four kaons are :

1. K−, negatively charged (containing a strange quark and an up antiquark) has mass 493.667±0.013 MeV and mean lifetime $(1.2384\pm0.0024)\times10^{-8}$ s.

2. K+ (antiparticle of above) positively charged (containing an up quark and a strange antiquark) must (by CPT invariance) have mass and lifetime equal to that of K−. The mass difference is 0.032±0.090 MeV, consistent with zero. The difference in lifetime is $(0.11\pm0.09)\times10^{-8}$ s.

3. K0, neutrally charged (containing a down quark and a strange antiquark) has mass 497.648±0.022 MeV. It has mean squared charge radius of −0.076±0.01 fm^2.

4. K0, neutrally charged (antiparticle of above) (containing a strange quark and a down antiquark) has the same mass.

It is clear from the quark model assignments that the kaons form two doublets of isospin; that is, they belong to the fundamental representation of SU(2) called the **2**. One doublet of strangeness +1 contains the K+ and the K0. The antiparticles form the other doublet (of strangeness −1).

[a] ^ Strong eigenstate. No definite lifetime (see kaon notes below)
[b] ^ Weak eigenstate. Makeup is missing small CP–violating term (see notes on neutral kaons below).
[c] ^ The mass of the K0
L and K0

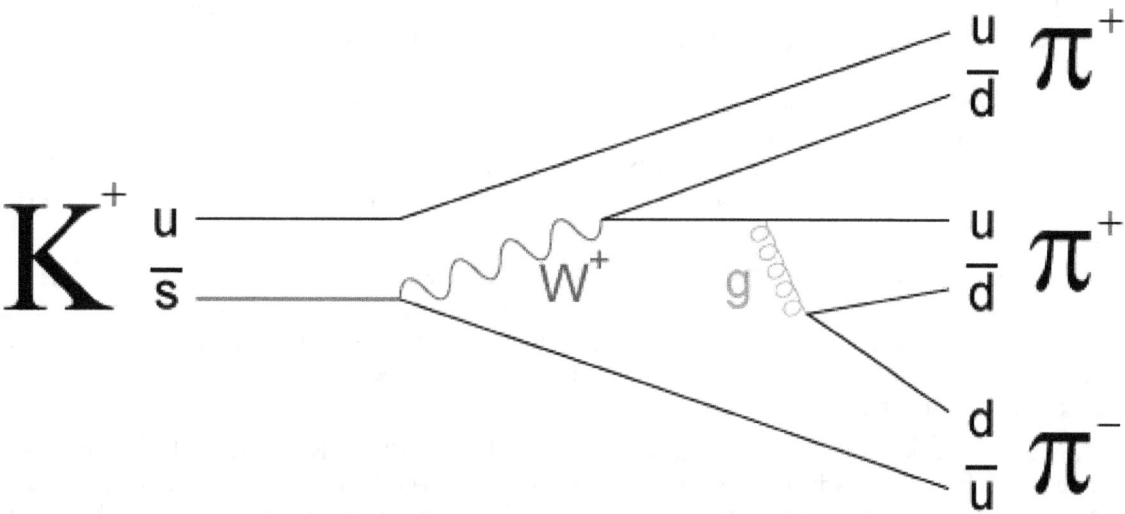

The decay of a kaon (K+) into three pions (2 π+, 1 π−) is a process that involves both weak and strong interactions.
Weak interactions : The strange antiquark (s) of the kaon transmutes into an up antiquark (u) by the emission of a W+ boson; the W+
boson subsequently decays into a down antiquark (d) and an up quark (u).
Strong interactions : An up quark (u) emits a gluon (g) which decays into a down quark (d) and a down antiquark (d).

S are given as that of the K0. However, it is known that a difference between the masses of the K0
L and K0
S on the order of 3.5×10^{-12} MeV/c^2 exists.[4]

Although the K0 and its antiparticle K0 are usually produced via the strong force, they decay weakly. Thus, once created the two are better thought of as superpositions of two weak eigenstates which have vastly different lifetimes:

1. The long-lived neutral kaon is called the K
   L ("K-long"), decays primarily into three pions, and has a mean lifetime of 5.18×10^{-8} s.

2. The short-lived neutral kaon is called the K
   S ("K-short"), decays primarily into two pions, and has a mean lifetime 8.958×10^{-11} s.

(See discussion of neutral kaon mixing below.)

An experimental observation made in 1964 that K-longs rarely decay into two pions was the discovery of CP violation (see below).

Main decay modes for K+:

Decay modes for the K− are charge conjugates of the ones above.

11.2   Strangeness

Main article: Strangeness

The discovery of hadrons with the internal quantum number "strangeness" marks the beginning of a most exciting epoch in particle physics that even now, fifty years later, has not yet found its conclusion ... by and large experiments have driven the development, and that major discoveries came unexpectedly or even against expectations expressed by theorists. — I.I. Bigi and A.I. Sanda, *CP violation*, (ISBN 0-521-44349-0)

In 1947, G. D. Rochester and Clifford Charles Butler of the University of Manchester published two cloud chamber photographs of cosmic ray-induced events, one showing what appeared to be a neutral particle decaying into two charged pions, and one which appeared to be a charged particle decaying into a charged pion and something neutral. The estimated mass of the new particles was very rough, about half a proton's mass. More examples of these "V-particles" were slow in coming.

The first breakthrough was obtained at Caltech, where a cloud chamber was taken up Mount Wilson, for greater cosmic ray exposure. In 1950, 30 charged and 4 neutral V-particles were reported. Inspired by this, numerous mountaintop observations were made over the next several years, and by 1953, the following terminology was adopted: "L-meson" meant muon or pion. "K meson" meant a particle intermediate in mass between the pion and nucleon. "Hyperon" meant any particle heavier than a nucleon.

The decays were extremely slow; typical lifetimes are of the order of 10^{-10} s. However, production in pion-proton reactions proceeds much faster, with a time scale of 10^{-23} s. The problem of this mismatch was solved by Abraham Pais who postulated the new quantum number called "strangeness" which is conserved in strong interactions but violated by the weak interactions. Strange particles appear copiously due to "associated production" of a strange and an antistrange particle together. It was soon shown that this could not be a multiplicative quantum number, because that would allow reactions which were never seen in the new synchrotrons which were commissioned in Brookhaven National Laboratory in 1953 and in the Lawrence Berkeley Laboratory in 1955.

11.3   Parity violation

Two different decays were found for charged strange mesons:

The intrinsic parity of a pion is $P = -1$, and parity is a multiplicative quantum number. Therefore, the two final states have different parity ($P = +1$ and $P = -1$, respectively). It was thought that the initial states should also have different parities, and hence be two distinct particles. However, with increasingly precise measurements, no difference was found between the masses and lifetimes of each, respectively, indicating that they are the same particle. This was known as the **τ–θ puzzle**. It was resolved only by the discovery of parity violation in weak interactions. Since the mesons decay through weak interactions, parity is not conserved, and the two decays are actually decays of the same particle,[5] now called the K+.

11.4   CP violation in neutral meson oscillations

Initially it was thought that although parity was violated, CP (charge parity) symmetry was conserved. In order to understand the discovery of CP violation, it is necessary to understand the mixing of neutral kaons; this phenomenon does not require CP violation, but it is the context in which CP violation was first observed.

11.4.1   Neutral kaon mixing

Since neutral kaons carry strangeness, they cannot be their own antiparticles. There must be then two different neutral kaons, differing by two units of strangeness. The question was then how to establish the presence of these two mesons. The solution used a phenomenon called **neutral particle oscillations**, by which these two kinds of mesons can turn from one into another through the weak interactions, which cause them to decay into pions (see the adjacent figure).

These oscillations were first investigated by Murray Gell-Mann and Abraham Pais together. They considered the CP-invariant time evolution of states with opposite strangeness. In matrix notation one can write

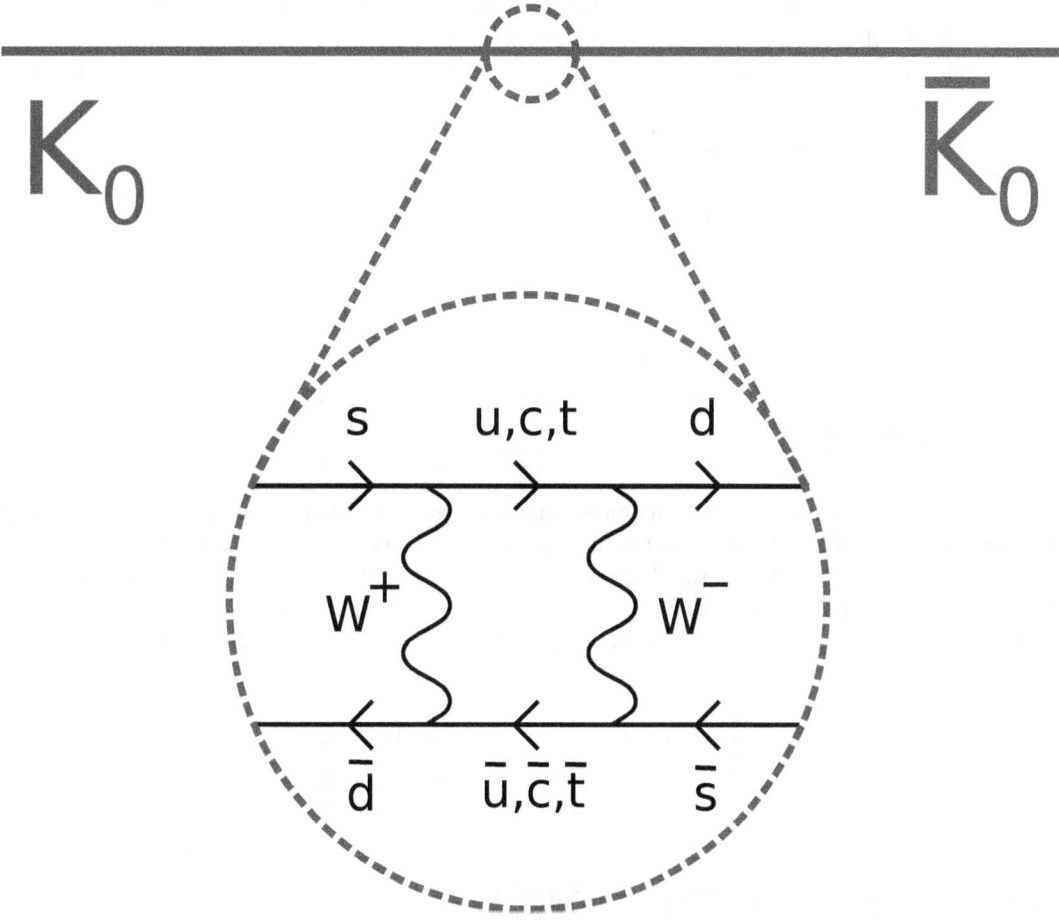

Two different neutral K mesons, carrying different strangeness, can turn from one into another through the weak interactions, since these interactions do not conserve strangeness. The strange quark in the K0 turns into a down quark by successively emitting two W-bosons of opposite charge. The down antiquark in the K0 turns into a strange antiquark by absorbing them.

$$\psi(t) = U(t)\psi(0) = e^{iHt}\begin{pmatrix} a \\ b \end{pmatrix}, \qquad H = \begin{pmatrix} M & \Delta \\ \Delta & M \end{pmatrix}$$

where ψ is a quantum state of the system specified by the amplitudes of being in each of the two basis states (which are a and b at time $t = 0$). The diagonal elements (M) of the Hamiltonian are due to strong interaction physics which conserves strangeness. The two diagonal elements must be equal, since the particle and antiparticle have equal masses in the absence of the weak interactions. The off-diagonal elements, which mix opposite strangeness particles, are due to weak interactions; CP symmetry requires them to be real.

The consequence of the matrix H being real is that the probabilities of the two states will forever oscillate back and forth. However, if any part of the matrix were imaginary, as is forbidden by CP symmetry, then part of the combination will diminish over time. The diminishing part can be either one component (a) or the other (b), or a mixture of the two.

Mixing

The eigenstates are obtained by diagonalizing this matrix. This gives new eigenvectors, which we can call $\mathbf{K}_1$ which is the difference of the two states of opposite strangeness, and $\mathbf{K}_2$, which is the sum. The two are eigenstates of **CP** with opposite eigenvalues; K_1 has $CP = +1$, and K_2 has $CP = -1$ Since the two-pion final state also has $CP = +1$, only the K_1 can decay this way. The K_2 must decay into three pions. Since the mass of K_2 is just a little larger than the sum of the masses of three pions, this decay proceeds very slowly, about 600 times slower than the decay of K_1 into two pions. These two different modes of decay were observed by Leon Lederman and his coworkers in 1956, establishing the existence of the two weak eigenstates (states with definite lifetimes under decays via the weak force) of the neutral kaons.

These two weak eigenstates are called the K
L (K-long) and K
S (K-short). CP symmetry, which was assumed at the time, implies that K
S = K_1 and K
L = K_2.

Oscillation

Main article: Neutral particle oscillation

An initially pure beam of K0 will turn into its antiparticle while propagating, which will turn back into the original particle, and so on. This is called particle oscillation. On observing the weak decay *into leptons*, it was found that a K0 always decayed into an electron, whereas the antiparticle K0 decayed into the positron. The earlier analysis yielded a relation between the rate of electron and positron production from sources of pure K0 and its antiparticle K0. Analysis of the time dependence of this semileptonic decay showed the phenomenon of oscillation, and allowed the extraction of the mass splitting between the K
S and K
L. Since this is due to weak interactions it is very small, 10^{-15} times the mass of each state.

Regeneration

A beam of neutral kaons decays in flight so that the short-lived K
S disappears, leaving a beam of pure long-lived K
L. If this beam is shot into matter, then the K0 and its antiparticle K0 interact differently with the nuclei. The K0 undergoes quasi-elastic scattering with nucleons, whereas its antiparticle can create hyperons. Due to the different interactions of the two components, quantum coherence between the two particles is lost. The emerging beam then contains different linear superpositions of the K0 and K0. Such a superposition is a mixture of K
L and K
S; the K
S is regenerated by passing a neutral kaon beam through matter. Regeneration was observed by Oreste Piccioni and his collaborators at Lawrence Berkeley National Laboratory. Soon thereafter, Robert Adair and his coworkers reported excess K
S regeneration, thus opening a new chapter in this history.

11.4.2 CP violation

While trying to verify Adair's results, J. Christenson, James Cronin, Val Fitch and Rene Turlay of Princeton University found decays of K
L into two pions ($CP = +1$) in an experiment performed in 1964 at the Alternating Gradient Synchrotron at the Brookhaven laboratory.[6] As explained in an earlier section, this required the assumed initial and final states to have different values of CP, and hence immediately suggested CP violation. Alternative explanations such as non-linear quantum mechanics and a new unobserved particle were soon ruled out, leaving CP violation as the only possibility. Cronin and Fitch received the Nobel Prize in Physics for this discovery in 1980.

It turns out that although the K
L and K

S are weak eigenstates (because they have definite lifetimes for decay by way of the weak force), they are *not quite* **CP** eigenstates. Instead, for small ε (and up to normalization),

K
L $= \mathbf{K}_2 + \varepsilon\mathbf{K}_1$

and similarly for K
S. Thus occasionally the K
L decays as a $\mathbf{K}_1$ with **CP** $= +1$, and likewise the K
S can decay with **CP** $= -1$. This is known as **indirect CP violation**, CP violation due to mixing of K0 and its antiparticle. There is also a **direct CP violation** effect, in which the CP violation occurs during the decay itself. Both are present, because both mixing and decay arise from the same interaction with the W boson and thus have CP violation predicted by the CKM matrix.

11.5 See also

- Hadrons, mesons, hyperons and flavour
- Strange quark and the quark model
- Parity (physics), charge conjugation, time reversal symmetry, CPT invariance and CP violation
- Neutrino oscillation
- Neutral particle oscillation

11.6 Notes and references

Notes

[1] The positively charged kaon used to be called τ^+ and θ^+, as it was supposed to be two different particles until the 1960s. See the parity violation section.

References

[1] J. Beringer *et al.* (2012): Particle listings – K±

[2] J. Beringer *et al.* (2012): Particle listings – K0

[3] J. Beringer *et al.* (2012): Particle listings – K0
S

[4] J. Beringer *et al.* (2012): Particle listings – K0
L

[5] Lee, T. D.; Yang, C. N. (1 October 1956). "Question of Parity Conservation in Weak Interactions". *Physical Review* **104** (1): 254. Bibcode:1956PhRv..104..254L. doi:10.1103/PhysRev.104.254. One way out of the difficulty is to assume that parity is not strictly conserved, so that Θ+ and τ+ are two different decay modes of the same particle, which necessarily has a single mass value and a single lifetime.

[6] http://journals.aps.org/prl/pdf/10.1103/PhysRevLett.13.138

11.6.1 Bibliography

- C.Amsler; Doser, M; Antonelli, M; Asner, D; Babu, K; Baer, H; Band, H; Barnett, R; Bergren, E; Bergren, E.; Beringer, J.; Bernardi, G.; Bertl, W.; Bichsel, H.; Biebel, O.; Bloch, P.; Blucher, E.; Blusk, S.; Cahn, R. N.; Carena, M.; Caso, C.; Ceccucci, A.; Chakraborty, D.; Chen, M.-C.; Chivukula, R. S.; Cowan, G.; Dahl, O.; d'Ambrosio, G.; Damour, T.; et al. (2008). "Review of Particle Physics". *Physics Letters B* (Particle Data Group) **667** (1): 1–1340. Bibcode:2008PhLB..667....1P. doi:10.1016/j.physletb.2008.07.018.

- S. Eidelman; et al. (2004). "Review of Particle Physics 2004 – Strange Mesons". Particle Data Group.

  Particle Data Group; Eidelman, S.; Hayes, K. G.; Olive, K. A.; Aguilar-Benitez, M.; Amsler, C.; Asner, D.; Babu, K. S.; Barnett, R. M.; Beringer, J.; Burchat, P. R.; Carone, C. D.; Caso, S.; Conforto, G.; Dahl, O.; d'Ambrosio, G.; Doser, M.; Feng, J. L.; Gherghetta, T.; Gibbons, L.; Goodman, M.; Grab, C.; Groom, D. E.; Gurtu, A.; Hagiwara, K.; Hernández-Rey, J. J.; Hikasa, K.; Honscheid, K.; Jawahery, H.; et al. (2004). "Review of Particle Physics*1". *Physics Letters B* **592** (1): 1. arXiv:astro-ph/0406663. Bibcode:2004PhLB..592....1P. doi:10.1016/j.physletb.2004.06.001.

- *The quark model*, by J.J.J. Kokkedee

- M.S. Sozzi (2008). *Discrete symmetries and CP violation*. Oxford University Press. ISBN 978-0-19-929666-8.

- I.I. Bigi, A.I. Sanda (2000). *CP violation*. Cambridge University Press. ISBN 0-521-44349-0.

- D.J. Griffiths (1987). *Introduction to Elementary Particle*. John Wiley & Sons. ISBN 0-471-60386-4.

Chapter 12

T meson

T mesons are hypothetical mesons composed of a top quark and either an up (T0), down (T+), strange (T+
s) or charm antiquark (T0
c).[1] Because of the top quark's short lifetime, T mesons are not expected to be found in nature. The combination of a top quark and top antiquark is not a T meson, but rather toponium. Each T meson has an antiparticle that is composed of a top antiquark and an up (T0), down (T−), strange (T−
s) or charm quark (T0
c) respectively.

12.1 References

[1] C. Amsler et al. (2008). "Review of Particle Physics: Naming Scheme for Hadrons" (PDF). *Physics Letters B* (Particle Data Group) **667** (1). Bibcode:2008PhLB..667....1P. doi:10.1016/j.physletb.2008.07.018.

12.2 External links

- W.-M. Yao *et al.* (Particle Data Group), J. Phys. G 33, 1 (2006) and 2007 partial update for edition 2008 (URL: http://pdg.lbl.gov)

Chapter 13

B meson

In particle physics, **B mesons** are mesons composed of a bottom antiquark and either an up (B+), down (B0), strange (B0 s) or charm quark (B+ c). The combination of a bottom antiquark and a top quark is not thought to be possible because of the top quark's short lifetime. The combination of a bottom antiquark and a bottom quark is not a B meson, but rather *bottomonium*.

Each B meson has an antiparticle that is composed of a bottom quark and an up (B−), down (B0), strange (B0 s) or charm antiquark (B− c) respectively.

13.1   List of B mesons

13.2   B–B oscillations

Main article: B–B oscillation

The neutral B mesons, B0 and B0 s, spontaneously transform into their own antiparticles and back. This phenomenon is called flavor oscillation. The existence of neutral B meson oscillations is a fundamental prediction of the Standard Model of particle physics. It has been measured in the B0–B0 system to be about 0.496 ps^{-1},[1] and in the B0 s–B0 s system to be $\Delta m_s = 17.77 \pm 0.10$ (stat) ± 0.07 (syst) ps^{-1} measured by CDF experiment at Fermilab.[2] A first estimation of the lower and upper limit of the B0 s–B0 s system value have been made by the DØ experiment also at Fermilab.[3]

On 25 September 2006, Fermilab announced that they had claimed discovery of previously-only-theorized B$_s$ meson oscillation.[4] According to Fermilab's press release:

> This first major discovery of Run 2 continues the tradition of particle physics discoveries at Fermilab, where the bottom (1977) and top (1995) quarks were discovered. Surprisingly, the bizarre behavior of the B_s (pronounced "B sub s") mesons is actually predicted by the Standard Model of fundamental particles and forces. The discovery of this oscillatory behavior is thus another reinforcement of the Standard Model's durability... CDF physicists have previously measured the rate of the matter-antimatter transitions for the B_s meson, which consists of the heavy bottom quark bound by the strong nuclear interaction to a strange

antiquark. Now they have achieved the standard for a discovery in the field of particle physics, where the probability for a false observation must be proven to be less than about 5 in 10 million (5/10,000,000). For CDF's result the probability is even smaller, at 8 in 100 million (8/100,000,000).

Ronald Kotulak, writing for the Chicago Tribune, called the particle "bizarre" and stated that the meson "may open the door to a new era of physics" with its proven interactions with the "spooky realm of antimatter".[5]

On 14 May 2010, physicists at the Fermi National Accelerator Laboratory reported that the oscillations decayed into matter 1% more often than into antimatter, which may help explain the abundance of matter over antimatter in the observed Universe.[6] However, more recent results at LHCb with larger data samples have suggested no significant deviation from the Standard Model.[7]

13.3   See also

- B–B oscillation

13.4   References

[1] http://repository.ubn.ru.nl/bitstream/2066/26242/

[2] A. Abulencia *et al.* (CDF Collaboration) (2006). "Observation of B0
     s–B0
     s Oscillations".*Physical Review Letters***97**(24): 242003.arXiv:hep-ex/0609040.Bibcode:2006PhRvL..97x2003A.doi:10.1103003.

[3] V.M. Abazov *et al.* (D0 Collaboration) (2006). "Direct Limits on the B_s^0 Oscillation Frequency" (PDF). *Physical Review Letters* **97** (2): 021802. arXiv:hep-ex/0603029. Bibcode:2006PhRvL..97b1802A. doi:10.1103/PhysRevLett.97.021802.

[4] "It might be...It could be...It is!!!" (Press release). Fermilab. 25 September 2006. Retrieved 2007-12-08.

[5] R. Kotulak (26 September 2006). "Antimatter discovery could alter physics: Particle tracked between real world, spooky realm". *Deseret News*. Archived from the original on 29 November 2007. Retrieved 2007-12-08.

[6] A New Clue to Explain Existence

[7] Article on LHCb results

13.5   External links

- W.-M. Yao *et al.* (Particle Data Group), J. Phys. G 33, 1 (2006) and 2007 partial update for edition 2008 (URL: http://pdg.lbl.gov)

- V. Jamieson (18 March 2008). "Flipping particle could explain missing antimatter". *New Scientist*. Retrieved 2010-01-23.

Chapter 14

Strange B meson

The **B**
s meson is a meson composed of a bottom antiquark and a strange quark. Its antiparticle is the **B**
s meson, composed of a bottom quark and a strange antiquark.

14.1   B–B oscillations

Strange B mesons are noted for their ability to oscillate between matter and antimatter via a box-diagram with $\Delta m_s = 17.77 \pm 0.10$ (stat) ± 0.07 (syst) ps^{-1} measured by CDF experiment at Fermilab.[1] That is, a meson composed of a bottom quark and strange antiquark, the strange B meson, can spontaneously change into an bottom antiquark and strange quark pair, the strange B meson, and vice versa.

On 25 September 2006, Fermilab announced that they had claimed discovery of previously-only-theorized B_s meson oscillation.[2] According to Fermilab's press release:

> This first major discovery of Run 2 continues the tradition of particle physics discoveries at Fermilab, where the bottom (1977) and top (1995) quarks were discovered. Surprisingly, the bizarre behavior of the B_s (pronounced "B sub s") mesons is actually predicted by the Standard Model of fundamental particles and forces. The discovery of this oscillatory behavior is thus another reinforcement of the Standard Model's durability...
>
> CDF physicists have previously measured the rate of the matter-antimatter transitions for the B_s meson, which consists of the heavy bottom quark bound by the strong nuclear interaction to a strange antiquark. Now they have achieved the standard for a discovery in the field of particle physics, where the probability for a false observation must be proven to be less than about 5 in 10 million (5/10,000,000). For CDF's result the probability is even smaller, at 8 in 100 million (8/100,000,000).[2]

Ronald Kotulak, writing for the Chicago Tribune, called the particle "bizarre" and stated that the meson "may open the door to a new era of physics" with its proven interactions with the "spooky realm of antimatter".[3]

Better understanding of the meson is one of the main objectives of the LHCb experiment conducted at the Large Hadron Collider.[4] On April 24, 2013, CERN physicists in the LHCb collaboration announced that they had observed CP violation in the decay of strange B mesons for the first time.[5][6] Scientists found the B_s meson decaying into two muons for the first time, with Large Hadron Collider experiments casting doubt on the scientific theory of supersymmetry.[7][8]

CERN physicist Tara Shears described the CP violation observations as "verification of the validity of the Standard Model of physics".[9]

14.2   Rare Decays

The rare decays of the B_s meson are an important test of the standard model. The branching fraction of the strange b-meson to a pair of muons is very precisely predicted with a value of $Br(B_s \rightarrow \mu^+\mu^-)SM = (3.66 \pm 0.23) \times 10^{-9}$. Any variation from this rate would indicate possible physics beyond the standard model, such as supersymmetry. The first definitive measurement was made from a combination of LHCb and CMS experiment data:[10]

$$Br(B_s \rightarrow \mu^+\mu^-) = 2.8^{+0.7}_{-0.6} \times 10^{-9}$$

This result is compatible with the standard model and set limits on possible extensions.

14.3   See also

- B meson

- Charmed B meson

- B–B oscillation

14.4   References

[1] A. Abulencia *et al.* (CDF Collaboration) (2006). "Observation of B0
    s–B0
    s Oscillations".*Physical Review Letters***97**: 242003.arXiv:hep-ex/0609040.Bibcode:2006PhRvL..97x2003A.doi:10.1103/Phys

[2] "It might be... It could be... It is!!!" (Press release). Fermilab. 25 September 2006. Retrieved 2007-12-08.

[3] R. Kotulak (26 September 2006). "Antimatter discovery could alter physics: Particle tracked between real world, spooky realm". *Deseret News*. Retrieved 2007-12-08.

[4] "A Taste of LHC Physics" (PDF). *Physics World*. June 2008. pp. 22–25.

[5] "LHCb experiment observes new matter-antimatter difference". CERN Press Office. 24 April 2013. Retrieved 2013-04-24.

[6]R. Aaij*et al.*(LHCb collaboration) (2013).*Physical Review Letters***110**(22): 221601.arXiv:1304.6173.Bibcode:2013PhRvL. doi:10.1103/PhysRevLett.110.221601. **Missing or empty |title= (help)

[7] M. Hogenboom (24 July 2013). "Ultra-rare decay confirmed in LHC". BBC. Retrieved 2013-08-18.

[8] CMS (14 May 2015). "Mathematical explanation from GENUINE published result". Nature. Retrieved 2015-05-15.

[9] M. Piesing (24 April 2013). "Cern physicists observe new difference between matter and antimatter". *Wired UK*. Retrieved 2013-04-24.

[10] Collaboration, C. M. S. (June 4, 2015). "Observation of the rare Bs0 →μ+μ– decay from the combined analysis of CMS and LHCb data". *Nature* **522** (7554): 68–72. doi:10.1038/nature14474. ISSN 0028-0836.

14.5   External links

- V. Jamieson (18 March 2008). "Flipping particle could explain missing antimatter". *New Scientist*. Retrieved 2010-01-23.

Chapter 15

Exotic meson

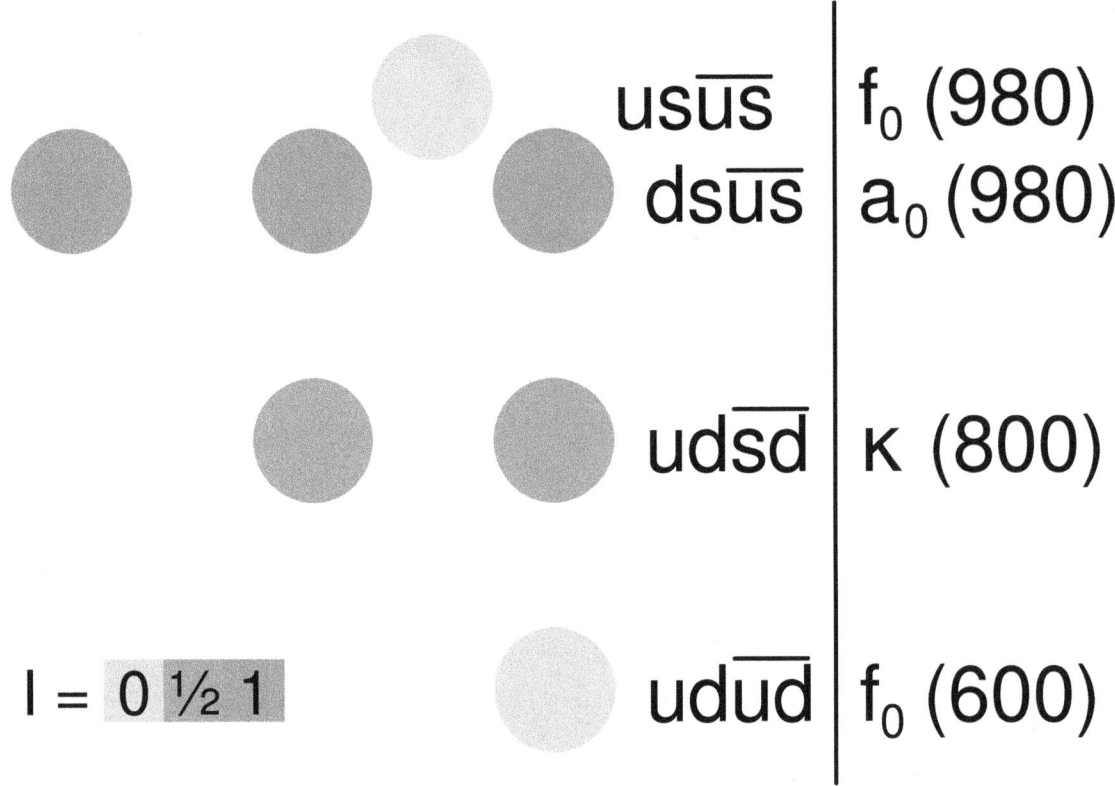

$us\overline{us}$ f_0 (980)

$ds\overline{us}$ a_0 (980)

$uds\overline{d}$ κ (800)

$ud\overline{ud}$ f_0 (600)

$I = $ 0 ½ 1

Identities and classification of possible tetraquark mesons. Green denotes I = 0 states, blue, I = 1/2 and red, I = 1. The vertical axis is the mass.

Non-quark model mesons include

1. **exotic mesons**, which have quantum numbers not possible for mesons in the quark model;

2. **glueballs** or **gluonium**, which have no valence quarks at all;

3. **tetraquarks**, which have two valence quark-antiquark pairs; and

4. **hybrid mesons**, which contain a valence quark-antiquark pair and one or more gluons.

All of these can be classed as mesons, because they are hadrons and carry zero baryon number. Of these, glueballs must be flavor singlets; that is, have zero isospin, strangeness, charm, bottomness, and topness. Like all particle states, they are specified by the quantum numbers which label representations of the Poincaré symmetry, q.e., J^{PC} (where J is the angular momentum, P is the intrinsic parity, and C is the charge conjugation parity) and by the mass. One also specifies the isospin I of the meson.

Typically, every quark model meson comes in SU(3) flavor nonet: an octet and a flavor singlet. A glueball shows up as an extra (*supernumerary*) particle outside the nonet. In spite of such seemingly simple counting, the assignment of any given state as a glueball, tetraquark, or hybrid remains tentative even today. Even when there is agreement that one of several states is one of these non-quark model mesons, the degree of mixing, and the precise assignment is fraught with uncertainties. There is also the considerable experimental labor of assigning quantum numbers to each state and cross-checking them in other experiments. As a result, all assignments outside the quark model are tentative. The remainder of this article outlines the situation as it stood at the end of 2004.

15.1   Lattice predictions

Lattice QCD predictions for glueballs are now fairly stable, at least when virtual quarks are neglected. The two lowest states are

0^{++} with mass of 1611 ± 163 MeV/c^2 and

2^{++} with mass of 2232 ± 310 MeV/c^2

The 0^{-+} and exotic glueballs such as 0^{--} are all expected to lie above 2 GeV/c^2. Glueballs are necessarily isoscalar, with isospin $I = 0$.

The ground state *hybrid mesons* 0^{-+}, 1^{-+}, 1^{--}, and 2^{-+} all lie a little below 2 GeV/c^2. The hybrid with exotic quantum numbers 1^{-+} is at 1.9 ± 0.2 GeV/c^2. The best lattice computations to date are made in the quenched approximation, which neglects virtual quarks loops. As a result, these computations miss mixing with meson states.

15.2   The 0^{++} states

The data show five isoscalar resonances: $f_0(500)$, $f_0(980)$, $f_0(1370)$, $f_0(1500)$, and $f_0(1710)$. Of these the $f_0(500)$ is usually identified with the σ of chiral models. The decays and production of $f_0(1710)$ give strong evidence that it is also a meson.

15.2.1   Glueball candidate

The $f_0(1370)$ and $f_0(1500)$ cannot both be a quark model meson, because one is supernumerary. The production of the higher mass state in two photon reactions such as $2\gamma \rightarrow 2\pi$ or $2\gamma \rightarrow 2K$ reactions is highly suppressed. The decays also give some evidence that one of these could be a glueball.

15.2.2   Tetraquark candidate

The $f_0(980)$ has been identified by some authors as a tetraquark meson, along with the $I = 1$ states $a_0(980)$ and $\kappa_0(800)$. Two long-lived (*narrow* in the jargon of particle spectroscopy) states: the scalar (0^{++}) state D*±
sJ(2317) and the vector (1^+) meson D*±
sJ(2460), observed at CLEO and BaBar, have also been tentatively identified as tetraquark states. However, for these, other explanations are possible.

15.3 The 2⁺⁺ states

Two isoscalar states are definitely identified—f$_2$(1270) and the f′$_2$(1525). No other states have been consistently identified by all experiments. Hence it is difficult to say more about these states.

15.4 The 1⁻⁺ exotics and other states

The two isovector exotics π_1(1400) and π_1(1600) seem to be well established experimentally. They are clearly not glueballs, but could be either a tetraquark or a hybrid. The evidence for such assignments is weak.

The π(1800) (0⁻⁺), ρ(1900) (1⁻⁻) and the η_2(1870) (2⁻⁺) are fairly well identified states, which have been tentatively identified as hybrids by some authors. If this identification is correct, then it is a remarkable agreement with lattice computations, which place several hybrids in this range of masses.

15.5 See also

- Quark model, mesons, baryons, quarks, and gluons

- Exotic hadrons and exotic baryons

- Quantum chromodynamics, flavor, and the QCD vacuum

- GlueX, an experiment which will explore the spectrum of glueballs and exotic mesons

15.6 References and external links

- W.-M. Yao *et al.* (Particle Data Group) (2006). "Review of Particle Physics: Non-qq mesons" (PDF). *Journal of Physics G* **33**: 1. arXiv:astro-ph/0601168. Bibcode:2006JPhG...33....1Y. doi:10.1088/0954-3899/33/1/001.

Chapter 16

Hadron spectroscopy

Hadron spectroscopy is the subfield of particle physics that studies the masses and decays of hadrons. Hadron spectroscopy is also an important part of the new nuclear physics. The properties of hadrons are a consequence of a theory called quantum chromodynamics (QCD).

QCD predicts that quarks and antiquarks bind into particles called mesons. Another type of hadron is called a baryon, that is made of three quarks. There is good experimental evidence for both mesons and baryons. Potentially QCD also has bound states of just gluons called glueballs. One of the goals of the field of hadronic spectroscopy is to find experimental evidence for exotic mesons, tetraquarks, molecules of hadrons, and glueballs.

An important part of the field of hadronic spectroscopy are the attempts to solve QCD. The properties of hadrons require the solution of QCD in the strong coupling regime, where perturbative techniques based on Feynman diagrams do not work. There are several approaches to trying to solve QCD to compute the masses of hadrons:

- Quark models

- Lattice QCD

- Effective field theory

- sum rules

16.1 Experimental facilities

- Jefferson Lab in the US.

- J-PARC in Japan.

- GSI Darmstadt Germany.

- COMPASS CERN, Switzerland.

16.2 References

- Article on Key Issues in Hadronic Physics

- Review of the quark model in PDG

16.3 Text and image sources, contributors, and licenses

16.3.1 Text

- **Meson** *Source:* https://en.wikipedia.org/wiki/Meson?oldid=672777660 *Contributors:* AxelBoldt, Bryan Derksen, Josh Grosse, PierreAbbat, Ben-Zin~enwiki, Xavic69, TakuyaMurata, Fwappler, Ahoerstemeier, Ping, Phys, Bcorr, Jeffq, Donarreiskoffer, Robbot, Fredrik, Sanders muc, Merovingian, Rursus, Ojigiri~enwiki, Davidl9999, DocWatson42, Harp, Marcika, Xerxes314, Niteowlneils, Eequor, Physicist, Eroica, Icairns, Sam Hocevar, Lehi, Rich Farmbrough, Pjacobi, Tjic, Robotje, Nicke Lilltroll~enwiki, Pearle, Jumbuck, Jérôme, Bucephalus, Falcorian, Palica, Tevatron~enwiki, Mandarax, Kbdank71, Strait, Titoxd, FlaBot, Jeremygbyrne, Chobot, YurikBot, Wavelength, Bambaiah, Phmer, Jimp, Ozabluda, JabberWok, Salsb, Leutha, Długosz, SCZenz, Ravedave, Gadget850, Antiduh, Tetracube, SmackBot, Melchoir, Eskimbot, Chris the speller, DHN-bot~enwiki, Sbharris, Kevinpurcell, Mesons, DMacks, Jashank, JorisvS, Mgiganteus1, Geologyguy, Ryulong, JarahE, Myasuda, ChrisKennedy, Michael C Price, Thijs!bot, Headbomb, Escarbot, Orionus, Spartaz, Gökhan, Deflective, Magioladitis, Swpb, Khalid Mahmood, Tercer, Kostisl, Hans Dunkelberg, Tarotcards, Xiahou, JeffreyRMiles, VolkovBot, Prizrak, TXiKiBoT, Muro de Aguas, Martin451, LeaveSleaves, Antixt, SieBot, Majeston, Gerakibot, Graf Von Crayola, Humanityisthedisease, Mimihitam, Fratrep, OKBot, ClueBot, Terrorist96, Diagramma Della Verita, Brews ohare, Neville35, RMFan1, WikHead, Stephen Poppitt, Addbot, Gtakanis, Chzz, Debresser, CosmiCarl, AgadaUrbanit, Dickdock, Magog the Ogre, AnomieBOT, StratoWiki, Altruism2010, Citation bot, ArthurBot, Xqbot, Omnipaedista, WaysToEscape, FrescoBot, Paine Ellsworth, Ironboy11, Steve Quinn, 000ojjo000, Yehoshua2, Citation bot 1, Wdcf, Thinking of England, Puzl bustr, Ale And Quail, Discovery4, Mean as custard, Dkzico007, John of Reading, WikitanvirBot, GoingBatty, Hanretty, ZéroBot, StringTheory11, Markinvancouver, ClueBot NG, Christian.kolen, Wallace Kneeland, Helpful Pixie Bot, Bibcode Bot, Glevum, DerekWinters, Mark viking, Justin567Hicks, Prokaryotes, SJ Defender, Monkbot, KasparBot and Anonymous: 87

- **Parity (physics)***Source:*https://en.wikipedia.org/wiki/Parity_(physics)?oldid=651198706*Contributors:*Patrick, TakuyaMurata, Charles Matthews,Phys, SoLando, Tobias Bergemann, Giftlite, Xerxes314, Beland, Karol Langner, Lumidek, CALR, Pak21, Nvj, Cmdrjameson, Eruantalon,Sergio Macías, Wtmitchell, Knowledge Seeker, Count Iblis, Oleg Alexandrov, Joriki, Marudubshinki, Ae77, Nihiltres, Thecurran, Wave-length, Bambaiah, Archelon, Pseudomonas, Kabirramola, E2mb0t~enwiki, Elkman, GrinBot~enwiki, SmackBot, Incnis Mrsi, Tom Lougheed,Leifisme, QFT, Wiki me, Akriasas, WhiteHatLurker, Erwin, JarahE, JRSpriggs, Raghunathan, Usgnus, Cydebot, Michael C Price, Thijs!bot,Barticus88, Mbell, Headbomb, Pjvpjv, Dougher, Magioladitis, Thasaidon, Dirac66, HEL, Tarotcards, Idioma-bot, Gerrit C. Groenenboom,Cuzkatzimhut, Red Act, Pamputt, Antixt, SieBot, BotMultichill, Paolo.dL, Anchor Link Bot, ClueBot, Sun Creator, DumZiBoT, Lazyrussian,Rror, TravisAF, Addbot, Luckas-bot, Yobot, Tonyrex, PianoDan, Citation bot, ArthurBot, Omnipaedista, Theaucitron, Sławomir Biały, CraigPemberton, Merongb10, RedBot, TobeBot, Linguisticgeek, Queller69, RjwilmsiBot, EmausBot, Albear-And, ZéroBot, Quondum, Kmva,ClueBot NG, Greedohun, Tamila Shalumova, Helpful Pixie Bot, Bibcode Bot, Vkpd11, Slumdog2011, Goodbear3, MuonRay, Abitslow, Jel-lyPatotie and Anonymous: 64

- **Isospin** *Source:* https://en.wikipedia.org/wiki/Isospin?oldid=682122579 *Contributors:* Stone, Giftlite, Xerxes314, Michael Devore, RScheiber, Jason Quinn, AmarChandra, Lumidek, Perey, Rich Farmbrough, Hidaspal, V79, Cmdrjameson, RJFJR, Linas, Robert K S, Jwanders, TPickup, Ddn2, FreplySpang, Rjwilmsi, Strait, Mike Peel, Margosbot~enwiki, Goudzovski, M7bot, Bambaiah, Bhny, Archelon, Welsh, Thiseye, SmackBot, Incnis Mrsi, Sue Anne, Colonies Chris, Sawran~enwiki, KI, Iridescent, Cydebot, Michael C Price, My Flatley, Zalgo, Thijs!bot, Headbomb, Knotwork, JAnDbot, Madmarigold, Avicennasis, Lilac Soul, KIAaze, Tarotcards, Fylwind, VolkovBot, Quilbert, Anonymous Dissident, Antixt, OlekG, PaddyLeahy, SieBot, Likebox, OsamaBinLogin, Uzdzislaw, Albambot, Addbot, Luckas-bot, Citation bot, ArthurBot, Bozzochet, Obersachsebot, Glenmark, Br77rino, J04n, Ernsts, RedAcer, Citation bot 1, Minivip, FoxBot, WikitanvirBot, Helpful Pixie Bot, Bibcode Bot, BG19bot, Jamisonsloan, Monkbot, Kfitzell29, GioComitini and Anonymous: 45

- **Flavour (particle physics)***Source:*https://en.wikipedia.org/wiki/Flavour_(particle_physics)?oldid=681888935*Contributors:*Schewek, MichaelHardy, Nurg, Xerxes314, Varlaam, Andycjp, R. fiend, DragonflySixtyseven, CALR, STGM, Andrew Gray, Knowledge Seeker, Egg, Alai,Sylvain Mielot, Linas, Mindmatrix, SpNeo, Drrngrvy, YurikBot, Bambaiah, Hairy Dude, NTBot~enwiki, Bhny, Cossy, Długosz, SCZenz,Nick, Karl Andrews, SmackBot, Incnis Mrsi, Dauto, Doug Bell, Zero sharp, Ompty, BFD1, Ruslik0, Cydebot, Hydraton31, Xxanthippe,Michael C Price, Thijs!bot, Headbomb, FelixP~enwiki, Rompe, Hayesgm, Knotwork, CosineKitty, Robin S, Askielboe, Yonidebot, Choihei,I310342~enwiki, Thecinimod, VolkovBot, A4bot, Kresadlo, Maxim, Odellus, Ptrslv72, SieBot, VVVBot, The Stickler, Muhends, PixelBot,Jtle515, Count Truthstein, DumZiBoT, MystBot, SkyLined, Addbot, ZeroOmega, SpBot, Ehrenkater, HerculeBot, Luckas-bot, Ptbotgourou,Magog the Ogre, Icalanise, Omnipaedista, Citation bot 1, Xtermin8R645, B2NVB2, Jrobbinz123, 777sms, Bizzurp, EmausBot, VinculumMan,AvocatoBot, Drift chambers, Skynden, Isambard Kingdom and Anonymous: 42

- **Strangeness** *Source:* https://en.wikipedia.org/wiki/Strangeness?oldid=674819737 *Contributors:* Xavic69, Ahoerstemeier, Timwi, Herbee, Xerxes314, JeffBobFrank, RScheiber, Icairns, Xeroc, Mike Rosoft, Jkl, Jag123, LostLeviathan, Fred Condo, Mel Etitis, Tevatron~enwiki, Eyu100, Donotresus, FlaBot, Who, Fresheneesz, Srleffler, Roboto de Ajvol, Bambaiah, Hairy Dude, Conscious, Shawn81, Kyorosuke, SCZenz, 99 Willys on Wheels on the wall, 99 Willys on Wheels..., SmackBot, Stepa, JSpudeman, Complexica, Richard L. Peterson, ZICO, ShelfSkewed, Cydebot, Dchristle, Mbell, Headbomb, AntiVandalBot, NE2, The sage, I310342~enwiki, Pernogr~enwiki, Anonymous Dissident, Pamputt, Riwnodennyk, Callie.hoon, SilvonenBot, Addbot, Mr0t1633, Zorrobot, Citation bot, Wnme, Ernsts, A. di M., Qwarx, Yutsi, Johann137, Turian, Alarichus, Dinamik-bot, EmausBot, Vacation9, 天才, Furkhaocean, JamesMoose, Ibnbaja and Anonymous: 33

- **Baryon number** *Source:* https://en.wikipedia.org/wiki/Baryon_number?oldid=675888853 *Contributors:* Andre Engels, Stevertigo, Delirium, Phys, Sanders muc, Securiger, Herbee, Xerxes314, Dratman, RScheiber, Jason Quinn, Discospinster, Pjacobi, Brim, Guy Harris, H2g2bob, Linas, Ted BJ, Isnow, Ddn2, BD2412, Raymond Hill, Bubba73, Margosbot~enwiki, Fresheneesz, Cannywizard, PointedEars, Roboto de Ajvol, YurikBot, Bambaiah, Tom Lougheed, V1adis1av, QFT, Doug Bell, Dan Gluck, Cydebot, Thijs!bot, Headbomb, Barakitty, Richard n, JAnDbot, CosineKitty, Bbi5291, Siryendor, FaTTshady74, STBotD, TXiKiBoT, A4bot, Venny85, SieBot, Muhends, Erodium, Addbot, WikiDreamer Bot, Legobot, Luckas-bot, Amirobot, ArthurBot, XZeroBot, Ernsts, MastiBot, Ttsush, Sahimrobot, Ernest3.141 and Anonymous: 17

- **Charm (quantum number)***Source:*https://en.wikipedia.org/wiki/Charm_(quantum_number)?oldid=669956667*Contributors:*Xavic69, Giftlite ,RScheiber, Pol098, Pdelong, Bambaiah, SmackBot, Tim Q. Wells, Cydebot, Thijs!bot, Headbomb, TXiKiBoT, Count Truthstein, ArthurBot, Ulm, Ernsts, Erik9bot, Carlog3, EmausBot, CocuBot, Ibnbaja, JuhoSchultz and Anonymous: 3

16.3.2 Images

- **File:Kaon-decay.png** *Source:* https://upload.wikimedia.org/wikipedia/commons/7/75/Kaon-decay.png *License:* CC-BY-SA-3.0 *Contributors:* Own work *Original artist:* User JabberWok on en.wikipedia

- **File:Meson_nonet_-_spin_0.svg** *Source:* https://upload.wikimedia.org/wikipedia/commons/c/c0/Meson_nonet_-_spin_0.svg *License:* Public domain *Contributors:* Image:Noneto mesônico de spin 0.png *Original artist:* User:E2m, User:Stannered

- **File:Meson_nonet_-_spin_1.svg** *Source:* https://upload.wikimedia.org/wikipedia/commons/1/13/Meson_nonet_-_spin_1.svg *License:* Public domain *Contributors:* Image:Noneto mesônico de spin 1.png *Original artist:* User:E2m, User:Stannered

- **File:Muon.svg** *Source:* https://upload.wikimedia.org/wikipedia/commons/d/d2/Muon.svg *License:* CC BY 3.0 *Contributors:* File:Standard Model of Elementary Particles.svg *Original artist:* user:MissMJ

- **File:Muon_neutrino.svg** *Source:* https://upload.wikimedia.org/wikipedia/commons/d/d0/Muon_neutrino.svg *License:* CC BY 3.0 *Contributors:* File:Standard Model of Elementary Particles.svg *Original artist:* user:MissMJ

- **File:Nuvola_apps_katomic.png** *Source:* https://upload.wikimedia.org/wikipedia/commons/7/73/Nuvola_apps_katomic.png *License:* LGPL *Contributors:* http://icon-king.com *Original artist:* David Vignoni / ICON KING

- **File:Office-book.svg** *Source:* https://upload.wikimedia.org/wikipedia/commons/a/a8/Office-book.svg *License:* Public domain *Contributors:* This and myself. *Original artist:* Chris Down/Tango project

- **File:Parity_1drep.png** *Source:* https://upload.wikimedia.org/wikipedia/commons/f/fc/Parity_1drep.png *License:* Public domain *Contributors:* ? *Original artist:* ?

- **File:Parity_clocks_-_P-conservation.svg** *Source:* https://upload.wikimedia.org/wikipedia/commons/6/65/Parity_clocks_-_P-conservation.svg *License:* CC BY-SA 3.0 *Contributors:* Own work *Original artist:* SiBr4

- **File:Parity_clocks_-_P-violation.svg** *Source:* https://upload.wikimedia.org/wikipedia/commons/1/1c/Parity_clocks_-_P-violation.svg *License:* CC BY-SA 3.0 *Contributors:* Own work *Original artist:* SiBr4

- **File:Portal-puzzle.svg** *Source:* https://upload.wikimedia.org/wikipedia/en/f/fd/Portal-puzzle.svg *License:* Public domain *Contributors:* ? *Original artist:* ?

- **File:Queryensdf.jpg** *Source:* https://upload.wikimedia.org/wikipedia/commons/5/5e/Queryensdf.jpg *License:* Public domain *Contributors:* Own work *Original artist:* Minivip

- **File:Question_book-new.svg** *Source:* https://upload.wikimedia.org/wikipedia/en/9/99/Question_book-new.svg *License:* Cc-by-sa-3.0 *Contributors:*
Created from scratch in Adobe Illustrator. Based on Image:Question book.png created by User:Equazcion *Original artist:*
Tkgd2007

- **File:Tau_lepton.svg** *Source:* https://upload.wikimedia.org/wikipedia/commons/f/f8/Tau_lepton.svg *License:* CC BY 3.0 *Contributors:* File:Standard Model of Elementary Particles.svg *Original artist:* user:MissMJ

- **File:Tau_neutrino.svg** *Source:* https://upload.wikimedia.org/wikipedia/commons/a/ac/Tau_neutrino.svg *License:* CC BY 3.0 *Contributors:* File:Standard Model of Elementary Particles.svg *Original artist:* user:MissMJ

- **File:Text_document_with_red_question_mark.svg** *Source:* https://upload.wikimedia.org/wikipedia/commons/a/a4/Text_document_with_red_question_mark.svg *License:* Public domain *Contributors:* Created by bdesham with Inkscape; based upon Text-x-generic.svg from the Tango project. *Original artist:* Benjamin D. Esham (bdesham)

16.3.3 Content license